Math Challenge I-A
Geometry

Areteem Institute

Math Challenge I-A Geometry

Edited by   John Lensmire
David Reynoso
Kevin Wang
Kelly Ren

Copyright © 2018 ARETEEM INSTITUTE

WWW.ARETEEM.ORG

PUBLISHED BY ARETEEM PRESS

ISBN: 1-944863-15-X
ISBN-13: 978-1-944863-15-9
First printing, August 2018.

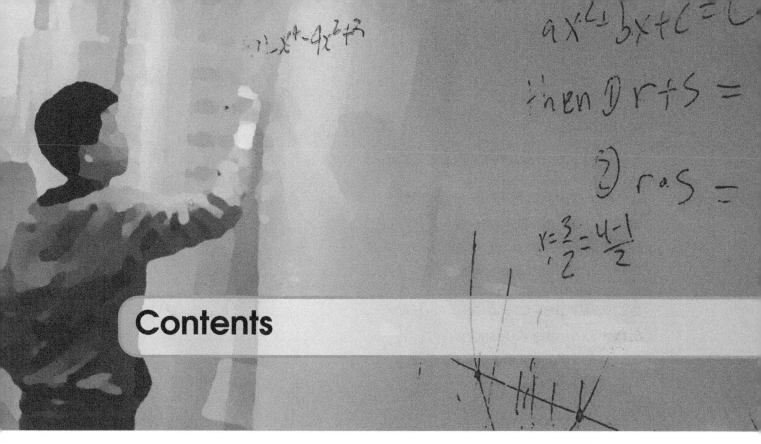

Contents

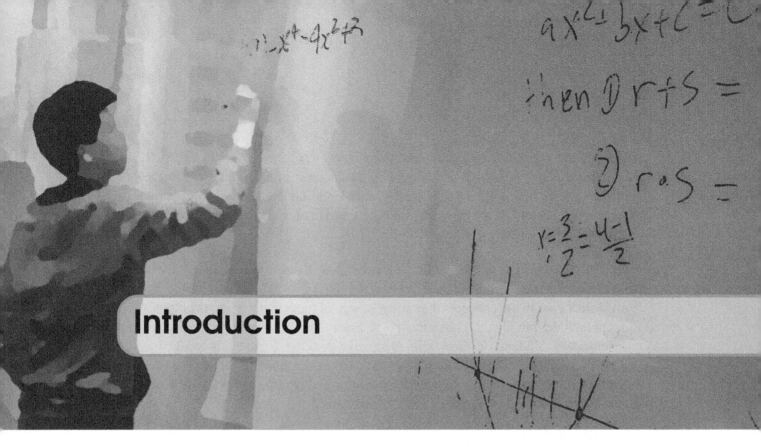

Introduction

The math challenge curriculum textbook series is designed to help students learn the fundamental mathematical concepts and practice their in-depth problem solving skills with selected exercise problems. Ideally, these textbooks are used together with Areteem Institute's corresponding courses, either taken as live classes or as self-paced classes. According to the experience levels of the students in mathematics, the following courses are offered:

- Fun Math Problem Solving for Elementary School (grades 3-5)
- Algebra Readiness (grade 5; preparing for middle school)
- Math Challenge I-A Series (grades 6-8; intro to problem solving)
- Math Challenge I-B Series (grades 6-8; intro to math contests e.g. AMC 8, ZIML Div M)
- Math Challenge I-C Series (grades 6-8; topics bridging middle and high schools)
- Math Challenge II-A Series (grades 9+ or younger students preparing for AMC 10)
- Math Challenge II-B Series (grades 9+ or younger students preparing for AMC 12)
- Math Challenge III Series (preparing for AIME, ZIML Varsity, or equivalent contests)
- Math Challenge IV Series (Math Olympiad level problem solving)

These courses are designed and developed by educational experts and industry professionals to bring real world applications into the STEM education. These programs are ideal for students who wish to win in Math Competitions (AMC, AIME, USAMO, IMO,

ARML, MathCounts, Math League, Math Olympiad, ZIML, etc.), Science Fairs (County Science Fairs, State Science Fairs, national programs like Intel Science and Engineering Fair, etc.) and Science Olympiad, or purely want to enrich their academic lives by taking more challenges and developing outstanding analytical, logical thinking and creative problem solving skills.

Math Challenge I-A is an introductory level course for 6-8 grade students who have little or no experience in in-depth problem solving nor math competitions. Students learn skills to apply the concepts they learn in school math classes into problem solving. Content includes pre-algebra, fundamental geometry, counting and probability, and basic number theory. Students develop skills in creative thinking, logical reasoning, analytical and problem solving skills. Students are exposed to beginning contests such as AMC 8, MathCounts, Math Olympiads for Elementary and Middle School (MOEMS), and Zoom International Math League (ZIML) Division M.

The course is divided into four terms:

- Summer, covering Pre-Algebra and Word Problems
- Fall, covering Geometry
- Winter, covering Counting and Probability
- Spring, covering Number Theory

The book contains course materials for Math Challenge I-A: Geometry.

We recommend that students take all four terms. Each of the individual terms is self-contained and does not depend on other terms, so they do not need to be taken in order, and students can take single terms if they want to focus on specific topics.

Students can sign up for the course at `classes.areteem.org` for the live online version or at `edurila.com` for the self-paced version.

About Areteem Institute

Areteem Institute is an educational institution that develops and provides in-depth and advanced math and science programs for K-12 (Elementary School, Middle School, and High School) students and teachers. Areteem programs are accredited supplementary programs by the Western Association of Schools and Colleges (WASC). Students may attend the Areteem Institute in one or more of the following options:

- Live and real-time face-to-face online classes with audio, video, interactive online whiteboard, and text chatting capabilities;
- Self-paced classes by watching the recordings of the live classes;
- Short video courses for trending math, science, technology, engineering, English, and social studies topics;
- Summer Intensive Camps held on prestigious university campuses and Winter Boot Camps;
- Practice with selected free daily problems and monthly ZIML competitions at ziml.areteem.org.

Areteem courses are designed and developed by educational experts and industry professionals to bring real world applications into STEM education. The programs are ideal for students who wish to build their mathematical strength in order to excel academically and eventually win in Math Competitions (AMC, AIME, USAMO, IMO, ARML, MathCounts, Math Olympiad, ZIML, and other math leagues and tournaments, etc.), Science Fairs (County Science Fairs, State Science Fairs, national programs like Intel Science and Engineering Fair, etc.) and Science Olympiads, or for students who purely want to enrich their academic lives by taking more challenging courses and developing outstanding analytical, logical, and creative problem solving skills.

Since 2004 Areteem Institute has been teaching with methodology that is highly promoted by the new Common Core State Standards: stressing the conceptual level understanding of the math concepts, problem solving techniques, and solving problems with real world applications. With the guidance from experienced and passionate professors, students are motivated to explore concepts deeper by identifying an interesting problem, researching it, analyzing it, and using a critical thinking approach to come up with multiple solutions.

Thousands of math students who have been trained at Areteem have achieved top honors and earned top awards in major national and international math competitions, including Gold Medalists in the International Math Olympiad (IMO), top winners and qualifiers at the USA Math Olympiad (USAMO/JMO) and AIME, top winners at the

Zoom International Math League (ZIML), and top winners at the MathCounts National Competition. Many Areteem Alumni have graduated from high school and gone on to enter their dream colleges such as MIT, Cal Tech, Harvard, Stanford, Yale, Princeton, U Penn, Harvey Mudd College, UC Berkeley, or UCLA. Those who have graduated from colleges are now playing important roles in their fields of endeavor.

Further information about Areteem Institute, as well as updates and errata of this book, can be found online at http://www.areteem.org.

Acknowledgments

This book contains many years of collaborative work by the staff of Areteem Institute. This book could not have existed without their efforts. Huge thanks go to the Areteem staff for their contributions!

The examples and problems in this book were either created by the Areteem staff or adapted from various sources, including other books and online resources. Especially, some good problems from previous math competitions and contests such as AMC, AIME, ARML, MATHCOUNTS, and ZIML are chosen as examples to illustrate concepts or problem-solving techniques. The original resources are credited whenever possible. However, it is not practical to list all such resources. We extend our gratitude to the original authors of all these resources.

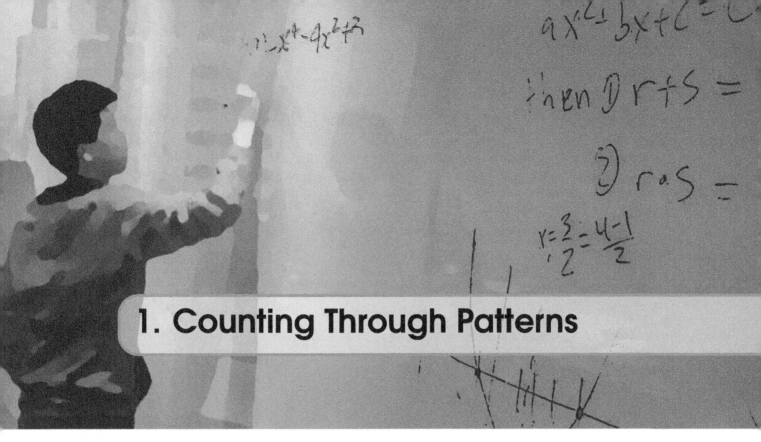

1. Counting Through Patterns

Fundamental Definitions in Geometry

- Point, Line, Shape
- Triangle, Rectangle, Parallelogram, Rhombus, Pentagon
- Intersection, Angle, Parallel
- Congruent, Equilateral, Equiangular

1.1 Example Questions

Problem 1.1 How many angles less than $180°$ are there in the following diagram?

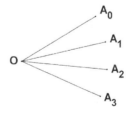

Problem 1.2 In the following diagram, $\angle 1 = \angle 2 = \angle 3$. The sum of the measures of all possible angles in $\angle AOB$ is $180°$. What is the measure of $\angle AOB$?

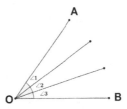

Problem 1.3 Count the triangles in each of the following diagrams.

(a)

(b)

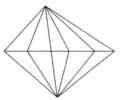

Problem 1.4 In the following diagram, each small equilateral triangle has area 1. Find the total area of all the triangles in the figure below.

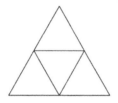

Problem 1.5 How many squares are there in the following diagram?

Problem 1.6 In the following diagram, $ABCD$ is a parallelogram, and each of the segments in the diagram is parallel to one of $\overline{AB}$, $\overline{AD}$, or $\overline{BE}$. Count the number of parallelograms in the diagram that contain the shaded triangle.

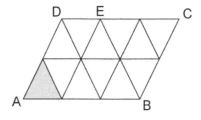

Problem 1.7 Arrange several equilateral triangles, all of whose side lengths are 2cm, to form a long parallelogram, as shown in the diagram. Assume the perimeter of the long parallelogram is 144cm, how many triangles are there?

Problem 1.8 Given a circular disk, use 3 lines to divide the disk into small regions. At most how many regions can there be? What if there are 4 lines?

Problem 1.9 In the diagram, each side is perpendicular to its adjacent sides, and all small sides have equal length. Given that the perimeter of this diagram is 108cm, find the area of the shape.

Problem 1.10 Use 4 congruent rectangles to form one big square, as shown. The big square has area 100 cm^2. Suppose the width of each rectangle is 1cm. What is the perimeter of each rectangle?

1.2 Quick Response Questions

Problem 1.11 How many points are there in the picture?

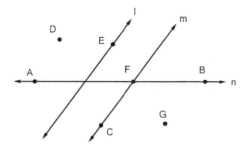

Problem 1.12 How many lines are there in the picture?

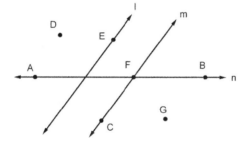

Problem 1.13 What points are on line *n*?

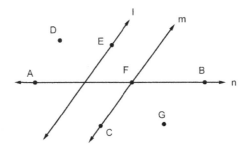

(A) *B,F,G*
(B) *A,B,F*
(C) *D,G*
(D) *C,F*

Problem 1.14 Which of the following points is not on any line?

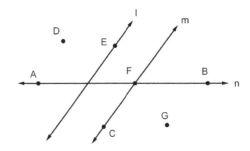

(A) *A*
(B) *B*
(C) *C*
(D) *D*

Problem 1.15 Which line is determined by the points C and F? That is, what is another name for the line $\overleftrightarrow{CF}$?

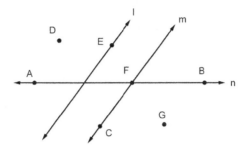

(A) l

(B) m

(C) n

(D) None of the above

Problem 1.16 Which of the following does not describe line n?

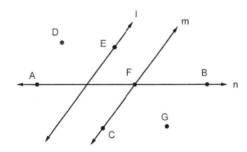

(A) $\overleftrightarrow{AB}$

(B) $\overleftrightarrow{AF}$

(C) $\overleftrightarrow{CF}$

(D) $\overleftrightarrow{BF}$

Problem 1.17 How many triangles are in the diagram below?

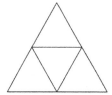

Problem 1.18 How many triangles are in the diagram below?

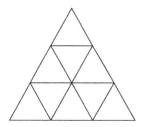

Problem 1.19 In the diagram, each side is perpendicular to its adjacent sides, and all small sides have equal length. Given that the perimeter of this diagram is 50, find the area of the shape.

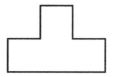

Problem 1.20 Arrange several equilateral triangles, all of whose side lengths are 1cm, to form a long parallelogram, as shown in the diagram. Assume the perimeter of the long parallelogram is 32cm, how many triangles are there?

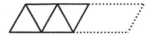

1.3 Practice Questions

Problem 1.21 How many angles less than $180°$ are there in the following diagram?

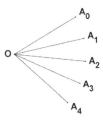

Problem 1.22 In the following diagram, $\angle 1 = 3\angle 3, \angle 2 = 2\angle 3$. The sum of the measures of all the angles is $180°$. What is the measure of $\angle AOB$?

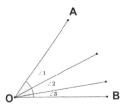

Problem 1.23 Count the triangles in each of the following diagrams.

(a)

(b)

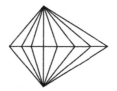

Problem 1.24 Each small equilateral triangle has area 1 in the diagram below. Find the total area of all the triangles.

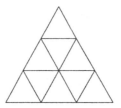

Problem 1.25 How many squares are there in the following diagram?

Problem 1.26 In the following diagram, $ABCD$ is a parallelogram, and each of the segments in the diagram is parallel to one of $\overline{AB}$, $\overline{AD}$, or $\overline{BE}$. Count the number of parallelograms in the diagram that contain the shaded triangle.

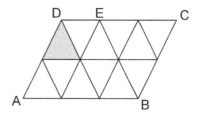

Problem 1.27 Arrange several rhombi, all of whose side lengths are 2cm, to form a long parallelogram, as shown in the diagram. Assume the perimeter of the long parallelogram is 144cm, how many rhombi are there?

Problem 1.28 Given a circular disk, use 6 lines to divide the disk into small regions. At most how many regions can there be?

Problem 1.29 In the diagram, each side is perpendicular to its adjacent sides, and all small sides have equal length. Given that the perimeter of this diagram is 108, find the area of the shape.

Problem 1.30 Use 4 congruent rectangles to form one big square, as shown. The big square has area 100 cm^2. Suppose the side length of the small square in the center is 2cm. What is the perimeter of each rectangle?

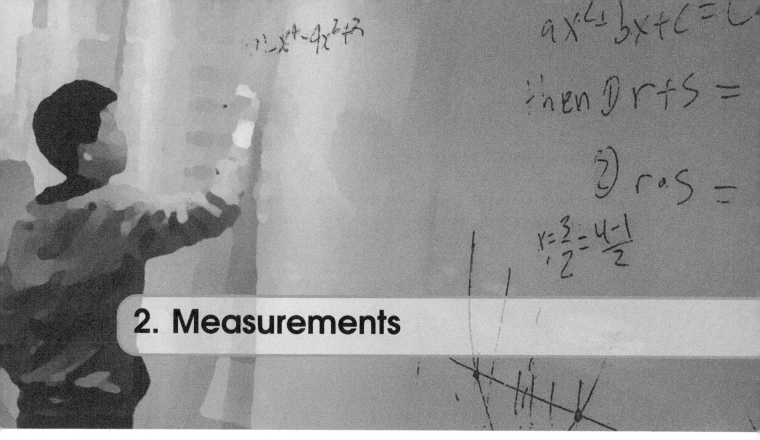

2. Measurements

Areas

- Area of a square of side s: $A = s^2$.
- Perimeter of a square of side s: $P = 4 \times s$.
- Area of a rectangle of length l and width w: $A = l \times w$.
- Perimeter of a rectangle of length l and width w: $P = 2 \times (l + w)$.

2.1   Example Questions

Problem 2.1 Two rectangles and two squares are assembled to form a big square as shown. The area of each rectangle is 28 and the area of the small square is 16. What is the area of the entire square?

Problem 2.2 A rectangle is divided into 3 squares, as shown in the diagram. Given that the area of one bigger square is 12 in^2 more than that of one smaller square, find the area of the whole rectangle.

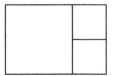

Problem 2.3 A rectangle is divided into 3 squares, as shown in the diagram. Given that the area of the rectangle is 150 in^2. Find the length and width of the rectangle.

Problem 2.4 A big rectangle is divided into 6 squares of different sizes, as shown. Given that the smallest square in the middle has area 4 cm^2 and the length of the big rectangle is 26, find the area of the big rectangle.

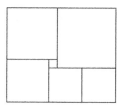

Problem 2.5 A big rectangle is divided into 7 smaller congruent rectangles, as shown. Given that the area of the big rectangle is 42 cm², find the perimeter of the big rectangle.

Problem 2.6 A rectangle is divided into 4 smaller rectangles by two lines, as shown. The perimeters of three of these rectangles are 12, 14, and 14. Find the perimeter of the remaining (shaded) rectangle.

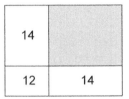

Problem 2.7 The perimeter of rectangle *ABCD* is 20 cm. Construct a square on the top and right sides of *ABCD* as shown below. Given that the sum of the areas of these squares is 52 cm^2, find the area of rectangle *ABCD*.

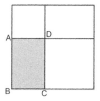

Problem 2.8 The shape in the diagram consists of 2 congruent squares and 3 congruent rectangles, and its perimeter is 14. Also given that $BC = \frac{1}{2}AB$. Find the area of rectangle *ABCD*.

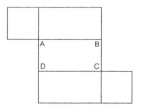

Problem 2.9 Four congruent rectangles and one square are assembled into one big square. The areas of the two squares are 144 and 16 respectively. What are the length and width of the rectangles?

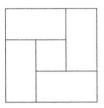

Problem 2.10 Divide a big square into 6 congruent rectangles, as shown. Given that each of the rectangles has perimeter 140, find the area of the big square.

2.2 Quick Response Questions

Problem 2.11 The following diagram is not to scale. The length of *AB* is 3 and the length of *AD* is 4.

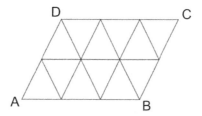

Find the perimeter of parallelogram *ABCD*.

Problem 2.12 The following diagram is not to scale. The length of *AB* is 3 and the length of *AD* is 4.

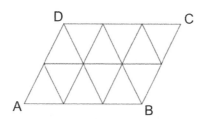

Find the perimeter of the one of the small triangles.

Problem 2.13 The following diagram is not to scale. The length of *AB* is 3 and the length of *AD* is 4.

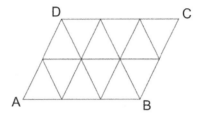

Suppose we remove the bases of all the small triangles and I want to travel on a zig-zag path from *A* to *B*. How long is the path I take?

Problem 2.14 In the following diagram the shaded square has area 5.

What is the area of the big square?

Problem 2.15 In the following diagram the shaded square has perimeter 8.

What is the area of the big square?

Problem 2.16 In the following diagram the largest triangle (made up with small triangles) has area 8.

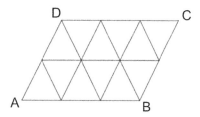

What is the area of a single small triangle?

Problem 2.17 In the following diagram the largest triangle (made up with small triangles) has area 8.

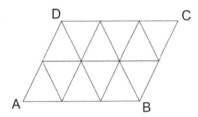

What is the area of the entire paralelogram?

Problem 2.18 Two rectangles and two squares are assembled to form a big square as shown. The area of each rectangle is 6 and the area of the small square is 4. What is the area of the entire square?

Problem 2.19 In the following diagram each of the small squares has perimeter 7.

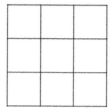

What is the perimeter of the big square?

Problem 2.20 In the following diagram the big square has perimeter 36.

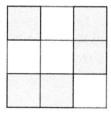

What is the area of the shaded region?

2.3 Practice Questions

Problem 2.21 Two rectangles and one square are assembled to form a big square as shown. The areas of the rectangles are 44 and 28. What is the area of the smaller (lower-right) square?

Problem 2.22 A rectangle is divided into 4 squares, as shown in the diagram. Given that the area of one bigger square is 16 in^2 more than that of one smaller square, find the area of the whole rectangle.

Problem 2.23 A rectangle is divided into 4 squares, as shown in the diagram. Given that the area of the rectangle is 300 in^2. Find the length and width of the rectangle.

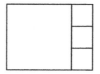

Problem 2.24 A big rectangle is divided into 6 squares of different sizes, as shown. Given the the smallest square in the middle has area 1 cm^2, find the area of the big rectangle.

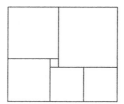

Problem 2.25 A big rectangle is divided into 10 smaller congruent rectangles, as shown. Given that the area of the big rectangle is $120\,\text{cm}^2$, find the perimeter of the big rectangle.

Problem 2.26 A rectangle is divided into 4 smaller rectangles by two lines, as shown. The areas of three of these rectangles are 6, 12, and 10. Find the area of the remaining (shaded) rectangle.

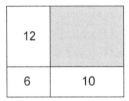

Problem 2.27 The perimeter of rectangle *ABCD* is 10 cm. Construct a square on the top and right sides of *ABCD* as shown below. Given that the sum of the areas of these squares is 13 cm^2, find the area of rectangle *ABCD*.

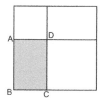

Problem 2.28 The shape in the diagram consists of 2 congruent squares and 3 congruent rectangles, and its perimeter is 28. Also given that $BC = \frac{1}{2}AB$. Find the total area of the diagram.

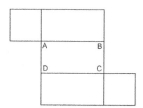

Problem 2.29 Four congruent rectangles and one square are assembled into one big square. The areas of the two squares are 64 and 16 respectively. What are the length and width of the rectangles?

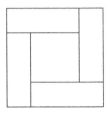

Problem 2.30 Divide a big square into 6 congruent rectangles, as shown. Given that each of the rectangles has perimeter 100, find the area of the big square.

3. Dance with Angles

Angles Review

- Angle: an *angle* is formed when two rays have a common origin. The common origin is the *vertex* of the angle, and the two rays are the *sides*.
- An entire circle is 360 degrees, or $360°$.
- A *right angle* is $90°$. Two lines (or rays or segments) that form a right angle are *perpendicular*. If $\angle BAC = 90°$, then $\overline{BA}$ and $\overline{AC}$ are perpendicular, or $\overline{BA} \perp \overline{AC}$.
- An angle less than $90°$ is called an *acute angle*.
- An angle greater than $90°$ but less than $180°$ is called an *obtuse angle*.
- A $180°$ angle is in fact a straight line, so it is called a *straight angle*.
- An angle greater than $180°$ is called a *reflex* angle.
- Two angles that add up to $180°$ are called *supplementary angles*.
- Two angles that add up to $90°$ are called *complementary angles*.

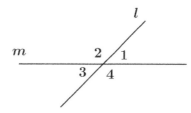

- In the above diagram,
    - *Adjacent Angles*: $\angle 1 + \angle 2 = 180°$
    - *Vertical Angles*: $\angle 1 = \angle 3$
  In summary: $\angle 1 = \angle 3$, and $\angle 2 = \angle 4$. Also, $\angle 1 + \angle 2 = \angle 3 + \angle 4 = 180°$.

Parallel Lines

- Let m and n be a pair of parallel lines (written $m\|n$). A third line l cuts across the parallel lines. Line l is called a *transversal*.

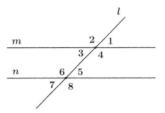

- In the above diagram,
    - *Corresponding angles*: $\angle 1 = \angle 5$
    - *Alternate interior angles*: $\angle 3 = \angle 5$
    - *Alternate exterior angles*: $\angle 2 = \angle 8$
    - *Same-side interior angles*: $\angle 4 + \angle 5 = 180°$
    - *Same-side exterior angles*: $\angle 1 + \angle 8 = 180°$

  In summary: $\angle 1 = \angle 3 = \angle 5 = \angle 7$, and $\angle 2 = \angle 4 = \angle 6 = \angle 8$. Also, $\angle 4 + \angle 5 = \angle 3 + \angle 6 = \angle 1 + \angle 8 = \angle 2 + \angle 7 = 180°$.

- In fact, if one the above equations hold, then m, n must be parallel. (And therefore all the equations hold.)

3.1 Example Questions

Problem 3.1 Consider the following diagram of two parallel lines and two transversals which meet at point C.

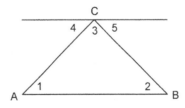

Which angles are equal in the above figure? Why?

Problem 3.2 Consider the previous diagram of two parallel lines and two transversals which meet at point C. What is $\angle 3 + \angle 4 + \angle 5$?

Problem 3.3 Consider the previous diagram of two parallel lines and two transversals which meet at point C. What is the sum of the angles in triangle ABC? Why?

Problem 3.4 Suppose that in the diagram below we have parallel lines and a transversal.

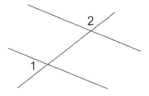

If the measure of $\angle 1$ is half the measure of $\angle 2$, find the measure of $\angle 1$.

Problem 3.5 Consider the diagram below:

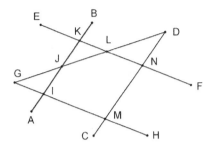

Suppose we know that $\overleftrightarrow{AB}$ and $\overleftrightarrow{CD}$ are parallel, $\angle DMH = 75°$, $\angle ELD = 140°$, and $\angle AJG = 35°$. What is $\angle LND$?

Problem 3.6 Consider the following "star" diagram, not drawn to scale.

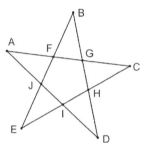

Suppose $\angle EFC = 120°$, $\angle AIC = 100°$, and $\angle BEC = 35°$. Calculate $\angle CAD$.

Problem 3.7 Given the figure below, if $\angle DFE = 60°$ and $\angle BCF = 90°$, what is the measure of $\angle CAF$?

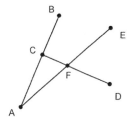

Problem 3.8 Suppose $\angle AOF = 180°$ and is divided into five equal angles as shown below.

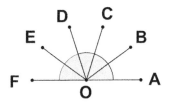

If $\angle AOB = \angle BOC = \cdots = \angle EOF$, find $\angle BOE$.

Problem 3.9 Consider the diagram below, where *l* and *m* are parallel but the drawing is not necessarily to scale.

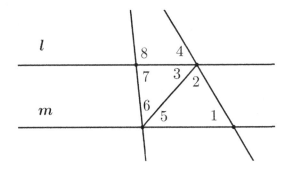

Suppose that $\angle 1 = 60°, \angle 5 = 50°, \angle 8 = 105°$. Find the measure of $\angle 4$.

Problem 3.10 Consider the diagram again.

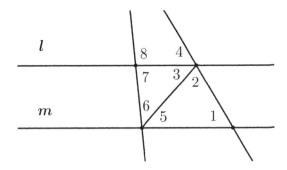

Suppose that $\angle 1 = 60°, \angle 5 = 50°, \angle 8 = 105°$. Find the measure of $\angle 2$.

3.2 Quick Response Questions

Problem 3.11 Classify the following angle.

(A) Acute
(B) Right
(C) Obtuse
(D) Reflex

Problem 3.12 Classify the following angle.

(A) Acute
(B) Right
(C) Obtuse
(D) Reflex

Problem 3.13 Classify the following angle.

(A) Acute
(B) Right
(C) Obtuse
(D) Reflex

Problem 3.14 Let m and n be a pair of parallel lines and let transversal l cut across the parallel lines as shown in the figure below.

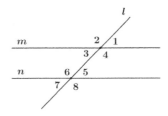

Which of the following angles are corresponding angles?

(A) $\angle 1 = \angle 3$
(B) $\angle 3 = \angle 7$
(C) $\angle 2 = \angle 8$
(D) $\angle 5 = \angle 7$

Problem 3.15 Let m and n be a pair of parallel lines and let transversal l cut across the parallel lines as shown in the figure below.

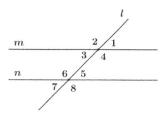

Which of the following are vertical angles?

(A) $\angle 1 = \angle 5$

(B) $\angle 2 = \angle 8$

(C) $\angle 5 = \angle 3$

(D) $\angle 6 = \angle 8$

Problem 3.16 Let m and n be a pair of parallel lines and let transversal l cut across the parallel lines as shown in the figure below.

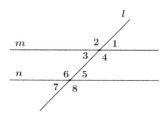

Which of the following are alternating interior angles?

(A) $\angle 3 = \angle 5$

(B) $\angle 1 = \angle 3$

(C) $\angle 5 = \angle 7$

(D) $\angle 3 = \angle 7$

Problem 3.17 Let m and n be a pair of parallel lines and let transversal l cut across the parallel lines as shown in the figure below.

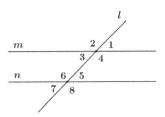

Which of the following are alternating exterior angles?

(A) $\angle 1 = \angle 5$
(B) $\angle 2 = \angle 6$
(C) $\angle 1 = \angle 3$
(D) $\angle 1 = \angle 7$

Problem 3.18 Let m and n be a pair of parallel lines and let transversal l cut across the parallel lines as shown in the figure below.

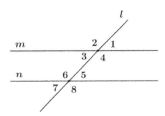

What kind of angles are $\angle 4$ and $\angle 5$?

(A) Alternate interior angles
(B) Alternate exterior angles
(C) Same-side interior angles
(D) Same-side exterior angles

Problem 3.19 Let m and n be a pair of parallel lines and let transversal l cut across the parallel lines as shown in the figure below.

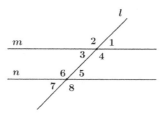

If $\angle 1 = 42°$, what is the measure of $\angle 7$?

Problem 3.20 Let m and n be a pair of parallel lines and let transversal l cut across the parallel lines as shown in the figure below.

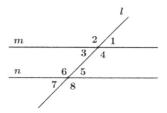

If $\angle 3 = 132°$, what is the measure of $\angle 5$?

3.3 Practice Questions

Problem 3.21 Consider the diagram below:

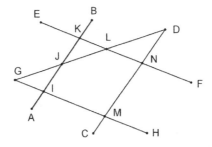

Suppose we know that $\overleftrightarrow{AB}$ and $\overleftrightarrow{CD}$ are parallel, $\angle DMH = 70°$, $\angle ELD = 135°$, and $\angle AJG = 30°$. What is $\angle LND$?

Problem 3.22 Consider the following "star" diagram, not drawn to scale.

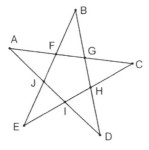

Suppose $\angle EFC = 115°$, $\angle AIC = 95°$, and $\angle BEC = 30°$. Calculate $\angle CAD$.

Problem 3.23 Suppose that in the diagram below we have parallel lines and a transversal.

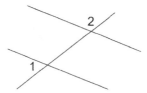

If the measure of $\angle 1$ is one-third the measure of $\angle 2$, find the measure of $\angle 1$.

Problem 3.24 Consider the diagram below:

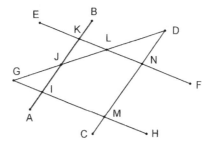

Suppose we know that $\overleftrightarrow{AB}$ and $\overleftrightarrow{CD}$ are parallel, $\angle DMH = 75°$, $\angle ELD = 140°$, and $\angle AJG = 35°$. What is $\angle BKE$?

Problem 3.25 Consider the following "star" diagram, not drawn to scale.

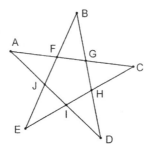

Suppose $\angle EFC = 120°$, $\angle AIC = 100°$, and $\angle BEC = 35°$. Calculate $\angle FJI$.

Problem 3.26 Given the figure below, if $\angle DFE = 75°$ and $\angle BCF = 95°$, what is the measure of $\angle CAF$?

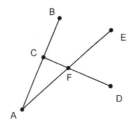

Problem 3.27 Suppose $\angle AOF = 180°$ and is divided into five equal angles as shown below.

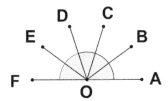

If $\angle AOB = \angle BOC = \cdots = \angle EOF$, find $\angle AOE$.

Problem 3.28 Using the above diagram, what is $\angle BOD$?

Problem 3.29 Consider the diagram below, where l and m are parallel but the drawing is not necessarily to scale.

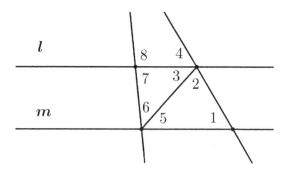

Suppose that $\angle 1 = 60°, \angle 5 = 50°, \angle 8 = 105°$. Find the measure of $\angle 6$.

Problem 3.30 Using the above diagram with the given measurements, find the measure of $\angle 3$.

Types of Triangles

- An *equilateral* triangle is a triangle made up of three equal sides and three equal angles.
- An *isosceles* triangle has two equal sides and the angles opposite those sides are equal.
- A *scalene* triangle has all different sides and all different angles.
- Note: In all of these examples, knowing about the sides OR the angles is enough. For example, if a triangle has two angles that are equal to each other, then the sides opposite those angles must be equal.

Congruent and Similar Triangles

- Two triangles are *congruent* if they are exactly the "same". More formally, this means all their sides and angles are equal.
- Two triangles are *similar* if they are the "same" except for possibly their size. More formally, this means their angles are equal and all their sides are in a common ratio (for example all the angles are the same but one triangle has sides that are all twice as long as the other triangle).
- In the diagram below $\triangle ABC$ is congruent to $\triangle DEF$ (written $\triangle ABC \cong \triangle DEF$). Triangle $\triangle GHI$ is NOT congruent to either $\triangle ABC$ or $\triangle DEF$. However, all three triangles are similar (written $\triangle ABC \sim \triangle DEF \sim \triangle GHI$).

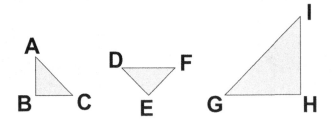

- If two triangles have all three sides the same length, then they are actually congruent. Think of making a triangle from three sticks, there is only one triangle you can make! This rule is often referred to as the **SSS** rule for congruence.
- Similarly, if two triangles are different sizes, but all the sides share a common ratio to each other they are similar. This rule is referred to as the **SSS** rule for similarity.

4.1 Example Questions

Problem 4.1 For each of the following "rules", state whether they work for proving congruence, similarity, both, or neither. If the rule does not work, give a counterexample.

(a) SAS (two sides and the angle between them)

(b) AAA (all three angles)

(c) ASA (two angles and the side between them)

(d) AAS (two angles and a side not between them)

(e) SSA (two sides and an angle not between them)

Problem 4.2 Suppose $\triangle ABC$ is an isosceles triangle with $\angle A = \angle B$. Let D be a point on $\overline{AB}$.

(a) Prove that D is the midpoint of $\overline{AB}$ (we call $\overline{CD}$ the *median* from C) if and only if $\angle ACD = \angle BCD$ (we call $\overline{CD}$ the *angle bisector* of $\angle ACB$).

(b) Further prove that $\overline{CD}$ is perpendicular to $\overline{AB}$ (we call $\overline{CD}$ the *altitude* from C).

Problem 4.3 Let $\overline{AD}$ and $\overline{BE}$ be medians in $\triangle ABC$. Prove that DE is half of AB.

Problem 4.4 Prove that the diagonals in a square are perpendicular.

Problem 4.5 Prove that in a quadrilateral $ABCD$, if $\overline{AB}$ is parallel to $\overline{CD}$ and $\overline{BC}$ is parallel to $\overline{AD}$, then $AB = CD$ (and similarly $BC = AD$).

Problem 4.6 Let D, E be midpoints of $\overline{AB}$ and $\overline{BC}$ in $\triangle ABC$. Let F be the intersection of the perpendicular bisectors of $\overline{AB}$ and $\overline{BC}$. Show that AF, BF, CF all have the same length.

Problem 4.7 Possible Triangles: Answer the following questions, with an explanation!

(a) Is it true that any three points are the vertices of some triangle?

(b) Given three numbers $p \leq q \leq r$ such that $p + q + r = 180$, is there a triangle $\triangle ABC$ such that $\angle A = p, \angle B = q, \angle C = r$? Hint: Suppose you make sure just $\angle A = p$ and $\angle B = q$, do you automatically know angle $\angle C = r$?

(c) Given three numbers $a \leq b \leq c$, is there a triangle $\triangle ABC$ with side lengths a, b, c? What can go wrong? Can you come up with a rule for when you can create a triangle with sides lengths a, b, c?

Problem 4.8 Suppose the two heights outside an obtuse triangle are the same length. Prove that the triangle is isosceles.

Problem 4.9 Suppose we have a "star" diagram as below (do not assume it is drawn to scale).

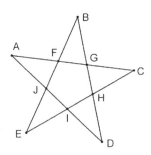

Now suppose that $\angle A = 30°$ and that $\triangle AFJ$ is isosceles. Calculate $\angle B + \angle D$.

Problem 4.10 Prove that if you connect the midpoints of the sides of an equilateral triangle it divides the triangle into four smaller congruent equilateral triangles.

4.2 **Quick Response Questions**

Problem 4.11 Is the following triangle an isosceles triangle?

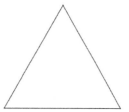

Problem 4.12 Is the following an isosceles right triangle?

Problem 4.13 Is the following an obtuse triangle?

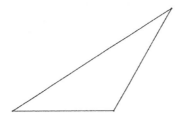

Problem 4.14 Is the following an obtuse scalene triangle?

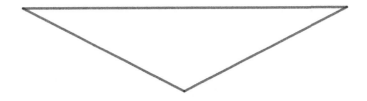

Problem 4.15 Is the following an isosceles right triangle?

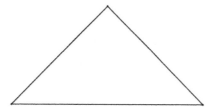

Problem 4.16 Is the following a scalene acute triangle?

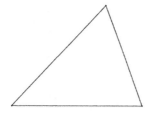

Problem 4.17 The following diagram is not to scale and lines that look paralel may not be.

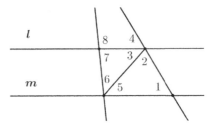

Suppose $\angle 1 = 45°$, $\angle 2 = 75°$, $\angle 6 = 60°$ and $\angle 7 = 70°$. Are l and m paralel?

Problem 4.18 Suppose an obtuse isosceles triangle has an angle of $30°$. What is the measure of its biggest angle?

Problem 4.19 Suppose an acute isosceles triangle has an angle of $20°$. What is the measure of its biggest angle?

Problem 4.20 Suppose a triangle has a $60°$ angle and the two sides adjacent to ir are equal. Is it an equilateral triangle?

4.3 Practice Questions

Problem 4.21 Prove that alternate exterior angles are equal and that same-side exterior angles are supplementary using the earlier results proved in class.

Problem 4.22 Prove that in a triangle an exterior angle is equal to the sum of the two interior angles not adjacent to it.

Problem 4.23 The star diagram from below may not be drawn to scale.

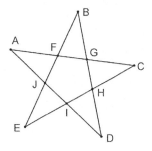

Suppose that $\angle A = 40°$ and that $\triangle AFJ$ is isosceles. Calculate $\angle B + \angle D$.

Problem 4.24 Prove that two angles are equal in a triangle if and only if the opposite sides are equal. (Recall that to prove an if and only if you need to prove both directions!)

Problem 4.25 Consider the diagram below.

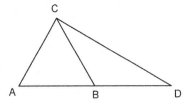

Suppose $AB = AC = 1$ and $\angle BAC = 60°$ and $\angle ADC = 30°$. Find BD.

Problem 4.26 Prove that if you connect the midpoints of the sides of an equilateral triangle it divides the triangle into four smaller congruent equilateral triangles.

Problem 4.27 Suppose in the diagram below that $\triangle ABC$ is isosceles and $\angle CAG = 20°$. Find the measure of $\angle ABD$.

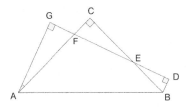

Problem 4.28 Prove that the diagonals of a parallelogram bisect each other.

Problem 4.29 Prove that the perpendicular bisectors of the three sides of a triangle $\triangle ABC$ all meet in a single point.

Problem 4.30 For each of the following, suppose you have $\triangle ABC$ satisfying the information given. Calculate as much of missing information as you can and draw the triangle.

(a) $a = b = c = 2$.

(b) $a = b = 4$ and $\angle C = 40°$.

(c) $\angle A = 50°, \angle C = 80°, a = 10$.

(d) $\angle A = 40°, \angle B = 70°, a = 10$.

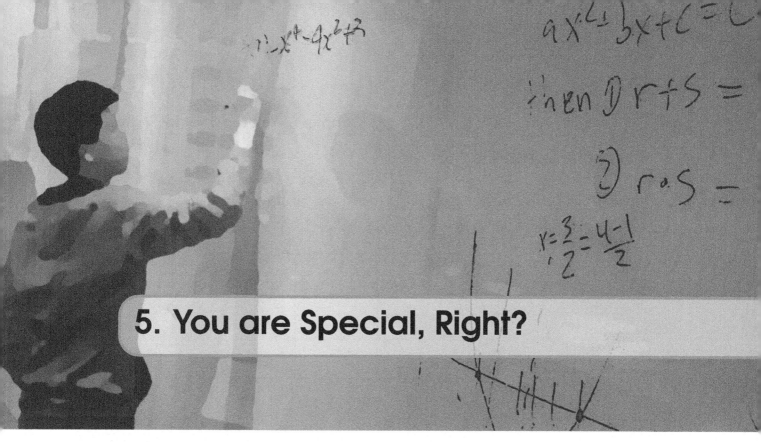

5. You are Special, Right?

Pythagorean Theorem

- In a right triangle $\triangle ABC$ with $\angle C = 90°$ and $AB = c, AC = b, BC = a$ we have $c^2 = a^2 + b^2$.
- Conversely, in a triangle $\triangle ABC$ with $AB = c, AC = b, BC = a$, if $c^2 = a^2 + b^2$ then $\triangle ABC$ is a right triangle.

Special Triangles

- An equilateral triangle has three $60°$ angles.
- In $\triangle ABC$, if $\angle C = 90°$, $\angle A = \angle B = 45°$, then $AB = \sqrt{2}AC = \sqrt{2}BC$.
- In $\triangle ABC$, if $\angle C = 90°$, $\angle A = 60°$, and $\angle B = 30°$, then $AB = 2AC$, $BC = \sqrt{3}AC$.

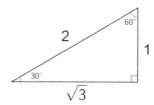
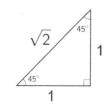

5.1 Example Questions

Problem 5.1 Prove the Pythagorean Theorem using the diagram below:

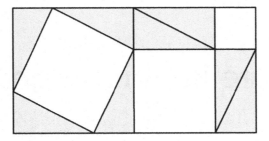

Problem 5.2 Prove the converse to the Pythagorean Theorem.

Problem 5.3 For each of the following, state whether it is possible to have a right triangle with the given side lengths. If it is possible, we call (a, b, c) a *Pythagorean Triple*.

(a) $3, 4, 5$.

(b) $4, 5, 6$.

(c) $5, 12, 13$.

(d) $6, 8, 10$.

(e) $5, 7, 8$.

Problem 5.4 Suppose you have a right triangle $\triangle ABC$ with hypotenuse $AC = 13$. Attach right triangle $\triangle BCD$ with hypotenuse BC to the side of $\triangle ABC$. If $\triangle BCD$ has sides of length $3, 4$, find AB.

Problem 5.5 Let $ABCD$ be a rectangle with $AB = 6, BC = 3$. Let E be the point a third of the way from A to B on $\overline{AB}$. Is $\angle CED$ a right angle?

Problem 5.6 Let's work with equilateral triangles.

(a) The sides of an equilateral triangle are 4 cm long. How long is an altitude of this triangle? What are the angles of a right triangle created by drawing an altitude? How does the short side of this right triangle compare with the other two sides?

(b) The altitudes of an equilateral triangle all have length 3 cm. How long are its sides?

(c) If the area of an equilateral triangle is $4\sqrt{3}$ square inches, how long are its sides?

Problem 5.7 Suppose that *ABCD* is a square. Let point *E* be *outside* the square and that △*CDE* is an equilateral triangle (see the diagram). What is the measure of ∠*EAD*?

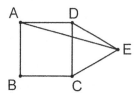

Problem 5.8 Given square *ABCD*, let *P* and *Q* be the points outside the square that make triangles *CDP* and *BCQ* equilateral. Segments $\overline{AQ}$ and $\overline{BP}$ intersect at *G*. Find angle *AGP*.

Problem 5.9 Given square *ABCD*, let *P* and *Q* be the points outside the square that make triangles *CDP* and *BCQ* equilateral. Prove that triangle *APQ* is also equilateral.

Problem 5.10 Draw the largest possible square inside an equilateral triangle, with one side of the square aligned with one side of the triangle. If the square has side length 6, find the side length of the equilateral triangle.

5.2 Quick Response Questions

Problem 5.11 If one of the leg length of a right triangle is 20 and the other leg length is 21, what is the length of the hypotenuse? Round your answer to the nearest hundredth if necessary.

Problem 5.12 If one of the leg length of a $45 - 45 - 90$ triangle is 3, what is the length of the hypotenuse? Round your answer to the nearest hundredth if necessary.

Problem 5.13 If the shorter leg length of a $30 - 60 - 90$ triangle is 3, what is the length of the hypotenuse? Round your answer to the nearest hundredth if necessary.

Problem 5.14 Consider the diagram below.

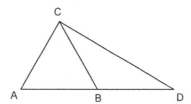

Suppose $AB = AC = 1$ and $\angle BAC = 60°$ and $\angle ADC = 30°$. Find BD.

Problem 5.15 The shorter leg of a right triangle is 6 and its hypotenuse is 10. What is the area of the triangle?

Problem 5.16 Let ABC be a right triangle with leg lengths 8 and 15. All lengths of ABC are doubled to form new triangle DEF. What is the area of triangle DEF?

Problem 5.17 The hypotenuse of an isosceles right triangle is $3\sqrt{2}$. What is the area of the triangle?

Problem 5.18 The larger smaller of a $30 - 60 - 90$ triangle is 8. Is it possible for its hypotenuse to be 10?

Problem 5.19 A right triangle has sides 8, $4\sqrt{3}$ and 4. What is the measure of the angle opposite to the side of length 4?

Problem 5.20 Which of the following is not a Pythagorean triple?

(A) $3,4,5$
(B) $4,5,6$
(C) $6,8,10$
(D) $5,12,13$

5.3   Practice Questions

Problem 5.21 Given two squares, show how to cut them up into pieces that can be combined to form one larger square. Your method should work no matter which two squares you are given. Hint: Start with the squares side by side as in the diagram below.

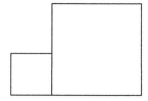

Problem 5.22 For each of the following, state whether it is possible to have a right triangle with the given side lengths. If it is possible, we call (a,b,c) a *Pythagorean Triple*.

(a) $6, 18, 21$

(b) $7, 24, 25$

(c) $8, 15, 17$

(d) $9, 18, 27$

(e) $10, 24, 26$

Problem 5.23 Suppose you have a right triangle $\triangle ABC$ with $AB = 8$ and $BC = 6$. Extend the line segment BC to point D so that the obtuse triangle $\triangle ACD$ is formed. If $CD = 9$, find the perimeter of $\triangle ACD$.

Problem 5.24 Let $ABCD$ be a rectangle with $AB = 18, BC = 5$. Let E be the point a third of the way from A to B on $\overline{AB}$. Is $\angle CED$ a right angle?

Problem 5.25 Let's work with isosceles right triangles.

(a) The legs of an isosceles right triangle are 4 cm long. How long is an altitude of this triangle? What are the angles of a right triangle created by drawing an altitude? How does the short side of this right triangle compare with the other two sides?

(b) The altitude of an isosceles right triangle that meets the hypotenuse of the triangle has length 3 cm. What is the length of the shortest side?

(c) If the area of an isosceles right triangle is 2 square inches, how long is the shortest side?

Problem 5.26 Suppose that $ABCD$ is a square. Let point E be inside the square and that $\triangle CDE$ is an equilateral triangle. What is the measure of $\angle EAD$?

Problem 5.27 Mark P inside square $ABCD$, so that triangle ABP is equilateral. Let Q be the intersection of BP with diagonal AC. Triangle CPQ looks isosceles. Is this actually true?

Problem 5.28 Given a parallelogram $ABCD$, let P and Q be the points outside the parallelogram so that triangles CDP and BCQ are equilateral. Is the triangle APQ is equilateral?

Problem 5.29 Draw the largest possible square inside an equilateral triangle, with one side of the square aligned with one side of the triangle. If the equilateral triangle has side length 6, find the side length of the square.

Problem 5.30 Suppose we have a "star" diagram as below (do not assume it is drawn to scale).

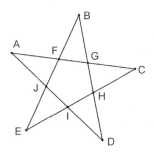

Now suppose that $\angle A = 40°$ and that $\triangle AFJ$ is isosceles. Calculate $\angle B + \angle D$.

6. Angles are Special Too

Special Angles

- Special angles are the angles $30°$, $60°$, $90°$ and $45°$.
- An equilateral triangle has three $60°$ angles.
- In $\triangle ABC$, if $\angle C = 90°$, $\angle A = \angle B = 45°$, then $AB = \sqrt{2}AC = \sqrt{2}BC$.
- In $\triangle ABC$, if $\angle C = 90°$, $\angle A = 60°$, and $\angle B = 30°$, then $AB = 2AC$, $BC = \sqrt{3}AC$.

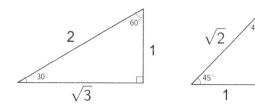

Polygons

- **Regular polygon:** A polygon that is both equilateral and equiangular is called regular.
- **Notation:** we use $[ABC]$ to denote the area of triangle ABC, $[DEFG]$ to denote the area of quadrilateral $DEFG$, etc.

Problem Solving Strategies

- Draw accurate diagrams using compass, ruler, and protractor.
- Try to find or make special angles in the diagram.

6.1 Example Questions

Problem 6.1 Complete the following table about polygons: name, sum of interior angles, sum of exterior angles, and measure of each angle in case of regular polygon. All angles are in degrees. Justify your answers. Keep the chart for your own reference.

# of sides	Polygon name	Int. angle sum	Ext. angle sum	Each angle (if regular)
3	Triangle			
4				
5				
6				
7	Heptagon			
8				
10				
12	Dodecagon			

Problem 6.2 Use 6 equilateral triangles to form a hexagon *ABCDEF*:

(a) Show hexagon *ABCDEF* is regular. Why is this true?

(b) Calculate the angle *AED*.

Problem 6.3 Three non-overlapping regular plane polygons all have sides of length 1. The polygons meet at a point *A* in such a way that the sum of the three interior angles at *A* is 360°. Among the three polygons, one is a triangle and one is a dodecagon. Find the remaining polygon.

Problem 6.4 Let *ABCDEF* be a regular hexagon, and let *EFGHI* be a regular pentagon. Find all possible values of measure of $\angle GAF$.

Problem 6.5 Suppose that *DRONE* is a regular pentagon, and that *DRUM*, *ROCK*, *ONLY*, *NEAP*, and *EDIT* are squares attached to the outside of the pentagon. Is the decagon *ITAPLYCKUM* equiangular? Is it equilateral?

Problem 6.6 Explain where and how to cut off the sides of an equilateral triangle to form a regular hexagon.

Problem 6.7 Practice with Hexagons

(a) Find the area of a regular hexagon with side length 12.

(b) Find the side length of a regular hexagon with area $150\sqrt{3}$.

Problem 6.8 Mark Y inside regular pentagon $PQRST$, so that PQY is equilateral. Is RYT straight? Explain.

Problem 6.9 Attach a regular pentagon $ABDEF$ to the side of an equilateral triangle ABC. Calculate $\angle CDE$.

Problem 6.10 A stop sign — a regular octagon — can be formed from a square sheet of metal by making four straight cuts that snip off the corners. If we want an octagon with sides of length $\sqrt{2}$, how large does the side of the original square need to be?

6.2 Quick Response Questions

Problem 6.11 Is the following a regular polygon?

Problem 6.12 Is the following an equiangular polygon that is not equilateral?

Problem 6.13 Is the following an equilateral polygon that is not equiangular?

Problem 6.14 Inside regular pentagon *JERZY* is marked point *P* so that triangle *JEP* is equilateral. Decide whether or not quadrilateral *JERP* is a parallelogram, and give your reasons.

Problem 6.15 A regular nonagon has nine sides. What is the the measure of each of its internal angles? Round your answer to the nearest hundredth if necessary.

Problem 6.16 A regular icosagon has twenty sides. What is the sum of the measure of its internal angles?

Problem 6.17 What is the sum of the internal angles of a rhombus?

Problem 6.18 A regular pentadecagon is a polygon with fifteen sides. What is the sum of its exterior angles?

Problem 6.19 An isosceles triangle has a repeated angle of 50°. What is the measure of its biggest angle?

Problem 6.20 Four of the angles of a pentagon add up to 410°. What is the measure of the fifth angle?

6.3   Practice Questions

Problem 6.21 Use 6 equilateral triangles to form a hexagon $ABCDEF$, then find the area of triangle AED in terms of the total area.

Problem 6.22 Equilateral triangles BCP and CDQ are attached to the outside of regular pentagon $ABCDE$. Is quadrilateral $BPQD$ a parallelogram? Justify your answer.

Problem 6.23 Three non-overlapping regular plane polygons all have sides of length 1. The polygons meet at a point A in such a way that the sum of the three interior angles at A is $360°$. Thus the three polygons form a new polygon P (not necessarily convex) with A as an interior point. Suppose two of the polygons are pentagons. Find the perimeter of P.

Problem 6.24 Let $ABCDEFGH$ be a regular octagon, and let $GHIJKL$ be a regular hexagon. Find all possible values of measure of $\angle IAH$.

Problem 6.25 Suppose that $DRONE$ is a regular pentagon, and that DRU, ROC, ONL, NEA, and EDI are equilateral triangles attached to the outside of the pentagon. Is $IUCLA$ a regular pentagon?

Problem 6.26 The equiangular convex hexagon $ABCDEF$ has $AB = 1$, $BC = 4$, $CD = 2$, and $DE = 4$. Find $[ABCDEF]$.

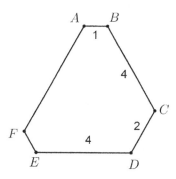

Problem 6.27 Practice with Octagons

(a) Find the area of a regular octagon with side length 12.

(b) Find the side length of a regular octagon with area $200 + 200\sqrt{2}$.

Problem 6.28 Mark Y inside regular hexagon $PQRSTU$, so that PQY is equilateral. Is RYU straight? Explain.

Problem 6.29 Suppose that $EARTH$ is a regular pentagon and regular pentagons $ANGLE$ and $RAPID$ are attached to the outside of the pentagon. Show that N, G, P, I are lie on a single line.

Problem 6.30 A stop sign — a regular octagon — can be formed from a square sheet of metal by making four straight cuts that snip off the corners. If we have a square with sides of length $\sqrt{2}$, what is the side length of the resulting octagon?

7. Area of Polygons

7.1 Example Questions

Problem 7.1 Using only the basics about parallel lines and congruent/similar triangles and the fact that the area of the rectangle is bh, explain the following: (Note: once a fact is proven below, you can use it in later parts.

(a) The area of a parallelogram is bh.

(b) The area of a triangle is $\frac{1}{2}bh$ (prove this two ways!).

(c) The area of a trapezoid is $\frac{b_1+b_2}{2}h$.

Problem 7.2 Explain the following:

(a) Triangles (or parallelograms) with equal bases and equal heights have the same area.

(b) In $\triangle ABC$, let D be the midpoint of side $\overline{BC}$ and connect $\overline{AD}$, as in the diagram below.

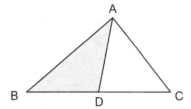

Prove that $[ABD] = [ACD]$ (that is, $\triangle ABD$ has the same area as $\triangle ACD$).

(c) Let $ABCD$ be a parallelogram and E be any point on side $\overline{CD}$ as shown.

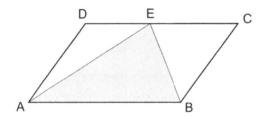

Prove that $[ABE] = \frac{1}{2}[ABCD]$.

(d) In $\triangle ABC$, let D be any point on side $\overline{BC}$ as shown below.

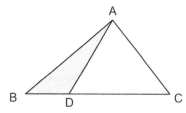

Prove that the ratios $\dfrac{[ABD]}{[ACD]} = \dfrac{BD}{CD}$.

Problem 7.3 Given three segments of lengths 4cm, 6cm, and 8cm. Use these lengths as the bases and altitude (not necessarily in the given order) to make a trapezoid. Trapezoids of three different possible areas can be made. Which one is the largest, and what is its area?

Problem 7.4 The parallelogram $ABCD$ has area 48cm^2, $AE = 8$cm, $CE = 4$cm. Find the area of the shaded region.

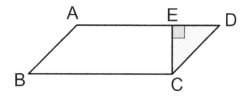

Problem 7.5 A garden of rectangular shape is shown in the diagram. The shaded regions are grass, and the unshaded regions are empty spaces in the shape of four congruent rhombi. Find the ratio between the areas of the grass and empty regions.

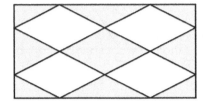

Problem 7.6 In the diagram, points A and B are the midpoints of their respective sides. Compute the ratio of the area of shaded region and the whole rectangle.

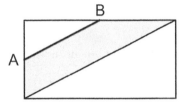

Problem 7.7 In the given trapezoid $ABCD$, there are 8 triangles. Among them, the pair $\triangle ABC$ and $\triangle DBC$ have the same area. How many other pairs have the same areas? List the pairs with the same areas.

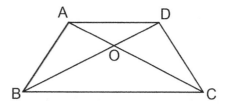

Problem 7.8 Let $ABCD$ be a parallelogram with $[ABCD]$. Let P be a point in the interior of $ABCD$. Show that $[ABP] + [CDP] = [ABCD]/2$.

Problem 7.9 Let $ABCD$ be a parallelogram, as in the diagram. Compare the shaded regions $\triangle ABF$ and $\triangle DEF$, which one has the larger area?

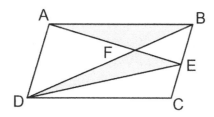

Problem 7.10 If regular hexagon $ABCDEF$ with side length 2 can be reinterpreted as 6 equilateral triangles, what is the area of $ABCDEF$?

7.2 Quick Response Questions

Problem 7.11 Let $ABCD$ be a square and let $EFGH$ be a rectangle such that $[ABCD] = [EFGH] = 16$. Suppose that the side length of the rectangle is twice the side length of the square. The ratio of the perimeter of the square to the perimeter of the rectangle is $a : b$, where a and b have no common factors. What is $a + b$?

Problem 7.12 What is the area of a regular hexagon with side length 3? Round your answer to the nearest tenth if necessary.

Problem 7.13 What is the area of a regular octagon with side length 3? Round your answer to the nearest tenth if necessary.

Problem 7.14 If the area of parallelogram $ABCD$ is 10 and the area of triangle CDE is 2, what is the area of trapezoid $ABCE$?

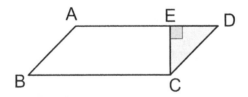

Problem 7.15 If the area of the rectangle is 40, the area of the smaller white triangle is 5, and the area of the big white triangle is 20, what is the area of the shaded region?

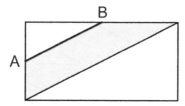

Problem 7.16 An equilateral triangle is attached to one of the sides of a regular hexagon as shown on the diagram.

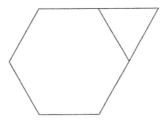

If the equilateral triangle has area 7, what is the area of the whole figure?

Problem 7.17 A garden of rectangular shape is shown in the diagram.

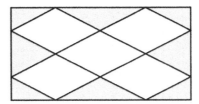

The shaded regions are grass. If the grass covers an area of 12 square feet, how many square feet is the whole garden?

Problem 7.18 A regular heptagon of area 84 is shown in the diagram below.

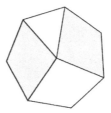

What is the shaded area?

Problem 7.19 In the following diagram point D is such that $AD : DB = 1 : 2$. If the area of the triangle is 144, what is the area of the shaded region?

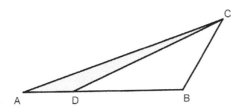

Problem 7.20 In the following diagram the shaded area is 50. What is the area of the entire regular pentagon?

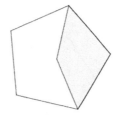

7.3 Practice Questions

Problem 7.21 Suppose you make a parallelogram with base 5cm and height 4cm.

(a) Find the area of the parallelogram.

(b) Show it is possible to have different perimeters for a parallelogram as above. Is there a largest perimeter that is possible? What about a smallest?

Problem 7.22 The parallelogram $ABCD$ has area 300cm^2 and E is the midpoint of $\overline{AD}$ Find the area of the shaded region.

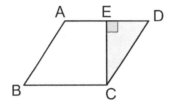

Problem 7.23 A garden of rectangular shape is shown in the diagram. The shaded regions are grass, and the unshaded regions are empty spaces in the shape of four congruent hexagons. Find the ratio between the areas of the grass and empty regions.

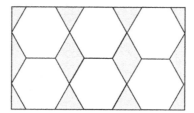

Problem 7.24 In the diagram, points A, B, C, D are the midpoints of their respective sides. Compute the ratio of the area of shaded region and the whole rectangle.

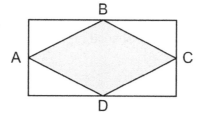

Problem 7.25 In the parallelogram $ABCD$, points E and F are the midpoints of sides $\overline{AD}$ and $\overline{DC}$ respectively. Which triangles have the same area as $\triangle BFC$?

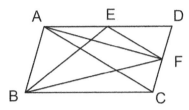

Problem 7.26 In the rectangle $ABCD$, the area of $\triangle AOB$ is 6 cm^2, and the area of $\triangle DOC$ is 1/3 of the area of the rectangle. Find the area of rectangle $ABCD$.

Problem 7.27 Let $ABCD$ be a parallelogram, as in the diagram. Suppose E is the midpoint of $\overline{BC}$. Find the area of the shaded regions $\triangle ABF$ and $\triangle DEF$ in terms of the entire area of the parallelogram.

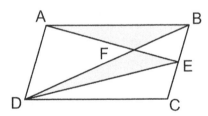

Problem 7.28 In the diagram, points *A* and *B* are the midpoints of their respective sides. Compute the ratio of the area of shaded region and the whole rectangle.

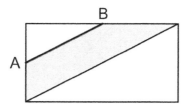

Problem 7.29 A garden of rectangular shape is shown in the diagram. The shaded regions are grass, and the unshaded regions are empty spaces in the shape of six congruent rhombi. Find the ratio between the areas of the grass and empty regions.

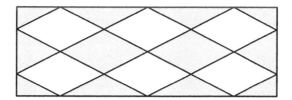

Problem 7.30 In the diagram, $\triangle ABC, \triangle DEF$ are two congruent isosceles right triangles. Given that $AB = 9, EC = 3$, find the area of the shaded region.

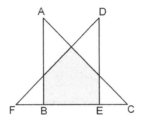

8. Cool Areas

8.1 Example Questions

Problem 8.1 All the rectangles in the following diagrams are squares. The lengths of the segments are marked. Find the area of the shaded regions in each diagram.

(a)

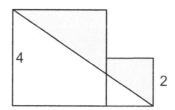

(b)

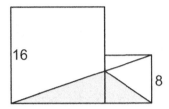

(c)

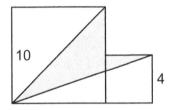

(d)

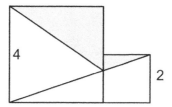

Problem 8.2 Suppose *ABCD* is a parallelogram, and the altitude $\overline{AE}$ on side $\overline{BC}$ is 5. Suppose the difference between the areas $[ADCE]$ and $[ABE]$ is 15. Find the length of $\overline{EC}$.

Problem 8.3 In the diagram below, there are 36 rectangular grid points, evenly spaced, and the distance between each pair of adjacent points is 1. Find the area of $\triangle ABC$.

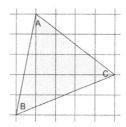

Problem 8.4 In the diagram below, there are 21 grid points arranged in equilateral triangles, equally spaced. The *area* of each small equilateral triangle formed by 3 adjacent grid points is 1. Find the area of $\triangle ABC$.

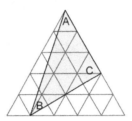

Problem 8.5 In the diagram, $\triangle ABC, \triangle DEF$ are two congruent isosceles right triangles. Given that $AB = 6, EC = 2$, find the area of the shaded region.

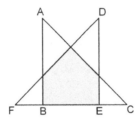

Problem 8.6 Suppose $\triangle ABC$ with E on $\overline{AB}$ and D on $\overline{AC}$ such that $AE = AB/3, AD = AC/2$. If $[AED] = 2$, find the area of $[ABC]$.

Problem 8.7 Parallelogram $ABCD$ is shown below, where triangle BCE is a right isosceles triangle and A is the midpoint of $\overline{BE}$. Given that $[ABCF] - [DFC] = 4$, find the area of $ABCD$.

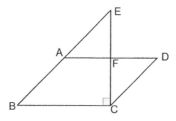

Problem 8.8 As shown in the diagram, square $ABCD$ has side length 5. Let E and F be the midpoints of $\overline{AB}$ and $\overline{BC}$ respectively. Find the area of quadrilateral $BFGE$.

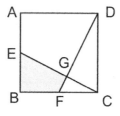

Problem 8.9 In the diagram, $\triangle ABC$, $\triangle DEF$ are two congruent isosceles right triangles. Given that $AB = 9, EC = 3$, find the area of the shaded region.

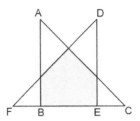

Problem 8.10 In the figure containing right triangle BCE and parallelogram $ABCD$ shown below, $EF = 3$, $EG = 5$ and $BG = 20$. Find the area of the shaded region.

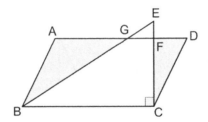

8.2 Quick Response Questions

Problem 8.11 Casey is building a tiny house. It will have a triangular roof that is 5m long (for each half), with a square bottom of the house. She plans to have the house be 8m wide. How tall is the house?

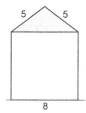

Hint: Do you remember some Pythagorean Triples?

Problem 8.12 Casey's friend Katherine is building a tiny house. It will have a triangular roof that is 5m long (for each half), with a square bottom of the house. She plans to have the house be 6m wide. How tall is the house?

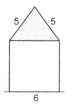

Problem 8.13 What is the amount of attic space in Casey's house?

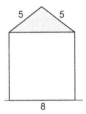

Problem 8.14 What is the total amount of space in Katherine's house (including the attic)?

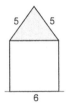

Problem 8.15 Suppose we have a right triangle such that $AC = 3$, $BC = 4$, and $\angle C = 90°$. What is the area of this triangle?

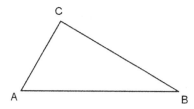

Problem 8.16 Suppose we have the a triangle such that $AC = 3$, $BC = 4$, and $\angle C = 90°$. The length of altitude CD is $\frac{P}{Q}$ in lowest terms. What is $P - Q$?

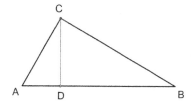

Problem 8.17 In the following diagram there are 10 grid points arranged in equilateral triangles, equaly spaced. The area of each small equilateral triangle formed by 3 adjacent grid points is 1. What is the area of the shaded region?

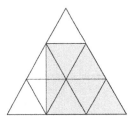

Problem 8.18 All the rectangles in the following diagrams are squares. What is the area of the shaded region?

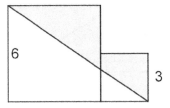

Problem 8.19 A rhombus is inscribed in a retangle as shown on the diagram below. If the area of the rectangle is 34, what is the shaded area?

Problem 8.20 On the following diagram the paralel sides are such that the bigger one is twice as long as the small one. If the shaded area is 17, what is the area of the trapezoid?

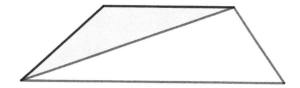

8.3 Practice Questions

Problem 8.21 All the rectangles in the following diagrams are squares. Find the area of the shaded regions in each diagram.

(a)

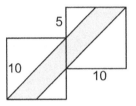

(b)

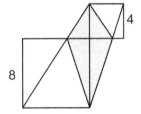

(c)

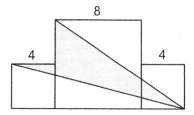

(d) The triangles below are right isosceles triangles.

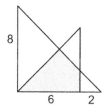

Problem 8.22 In parallelogram *ABCD*, *M* and *N* are midpoints of sides $\overline{AB}$ and $\overline{BC}$ respectively. Given that $[DMN] = 9$, find the area of *ABCD*.

Problem 8.23 In the diagram below, there are 36 rectangular grid points, evenly spaced, and the distance between each pair of adjacent points is 1. Find the area of quadrilateral *ABCD*.

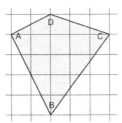

Problem 8.24 In the diagram below, there are 21 grid points arranged in equilateral triangles, equally spaced. The *area* of each small equilateral triangle formed by 3 adjacent grid points is 1. Find the area of quadrilateral *ABCD*.

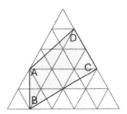

Problem 8.25 In the diagram, $\triangle ABC, \triangle DEF$ are two congruent isosceles right triangles. Given that $ADFC$ is a 4×2 rectangle, find the area of the shaded region.

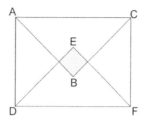

Problem 8.26 Suppose $\triangle ABC$ with D on $\overline{AB}$, G on $\overline{AC}$, and E, F on $\overline{BC}$. Suppose further $AD = AB/2, BE = EF = FC, CG = GA/2$. If the area of $[ABC] = 36$, find the area of the quadrilateral $DEFG$.

Problem 8.27 In the square shown in the diagram, the side length is 6, and the sum of the areas of the two shaded regions is 12. Find the area of quadrilateral $ABCD$.

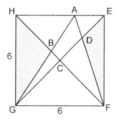

Problem 8.28 A square is formed by putting 4 congruent isosceles right triangles at the corners. The shaded square is the region not covered by the triangles. Find the area of the shaded square, if

(a) As shown in the diagram below, the triangles just touch.

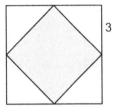

(b) As shown in the diagram below, the triangles overlap a little. (Note: In this case, 3 is *not* the side length of the isosceles triangle.)

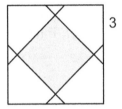

Problem 8.29 In parallelogram $ABCD$ as shown, $BC = 10$. Triangle BCE is a right triangle where $\overline{BE}$ is the hypotenuse, and $EC = 8$. Given that $[ABG] + [CDF] - [EFG] = 10$, find the length of $\overline{CF}$.

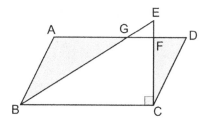

Problem 8.30 If $BE = 10$, $AD = 8$ and the area of right triangle BCE is equal to the area of parallelogram $ABCD$ in the diagram below,

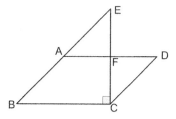

what is the area of trapezoid $ABCF$?

Basic Definitions

- A *circle* is a collection of points of equal distance (called the *radius*) from a set point (called the *center*).
- Given two points A, B on a circle, the segment $\overline{AB}$ is called a *chord*.
- If a chord AB contains the center of the circle, call AB a *diameter*.
- The portion of a circle that lies above or below a chord AB is called an *arc*. If the arc is more than half a circle it is called a *major arc*, less than half a circle is called a *minor arc*, and half a circle is called a *semicircle*. The arc will be denoted $\overset{\frown}{AB}$.
- Suppose $\overset{\frown}{AB}$ is an arc on a circle with center O. The *angular size* of the arc $\overset{\frown}{AB}$ is equal to the angle $\angle AOB$ (which is referred to as a *central angle*).
- Note: a full circle is $360°$, a half circle is $180°$, and a quarter circle is $90°$.
- Given a central angle $\angle AOB$ from an arc $\overset{\frown}{AB}$, the figure contained between the arc $\overset{\frown}{AB}$ and the radii $\overline{OA}, \overline{OB}$ is called a *sector*.
- The diagram below has a few examples:

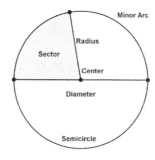

Measurements in Circles

- The area of a circle is given by πr^2 where r is the radius.
- The circumference of a circle is given by $2\pi r = \pi d$ where r, d are the radius and length of a diameter respectively.
- The area of a sector from arc $\overset{\frown}{AB}$ is given by $\dfrac{\theta}{360°}\pi r^2$ where θ is the angular size of $\overset{\frown}{AB}$ (measured in degrees).
- In particular, the area of half a circle is $\frac{180°}{360°}\pi r^2 = \frac{1}{2}\pi r^2$ and the area of a quarter circle $\frac{90°}{360°}\pi r^2 = \frac{1}{4}\pi r^2$.
- Similarly, *arc length* of $\overset{\frown}{AB}$ (that is, the distance walking from A to B along the circle) is given by $\dfrac{\theta}{360°}2\pi r$ where θ is the angular size of $\overset{\frown}{AB}$ (measured in degrees).
- In particular, the arc length of a semicircle is $\frac{180°}{360°}2\pi r = \frac{1}{2}2\pi r = \pi r$ and the arc length of a quarter circle is $\frac{90°}{360°}2\pi r = \frac{1}{4}2\pi r = \frac{1}{2}\pi r$.

Tangents

- A line is *tangent* to a circle is if the line intersects the circle exactly once. If this intersection point is P, we say the line is *tangent to the circle at P*.
- Similarly, two circles are tangent if they intersect at exactly one point. As before, if this point is P, we will say the two circles are tangent at a point A.
- Note: If two objects are tangent at a point P, it is often useful to think of them as "just touching" at P.
- **Fact**: If a line is tangent to a circle at point P, and $\overline{OP}$ is a radius of the circle. Then $\overline{OP}$ is perpendicular to the line, as in the picture below:

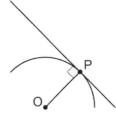

- **Fact**: If two circles (with centers N, O) are tangent at a point P, then the line perpendicular to $\overline{NO}$ going through P is tangent to both circles, as in the picture below:

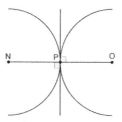

9.1 Example Questions

Problem 9.1 A circular dining table has diameter 2 meters and height 1 meter. A square tablecloth is placed on the table, and the four corners of the tablecloth just touch the floor. Find the area of the tablecloth in square meters.

Problem 9.2 The shape in the diagram consists of three semicircles with radii 5, 3, and 2, arranged as shown. Calculate the perimeter and area of the shaded region.

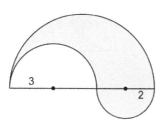

Problem 9.3 Find the areas of the shaded regions in each of the diagrams. Note: for the most part things are drawn to scale, so you may assume angles that look right are in fact 90°, etc.

(a)

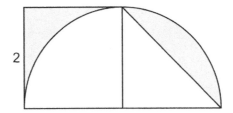

(b)

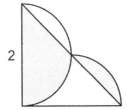

Problem 9.4 Find the areas of the shaded regions in each of the diagrams. Note: for the most part things are drawn to scale, so you may assume angles that look right are in fact 90°, etc.

(a)

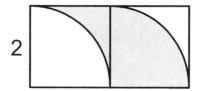

(b)

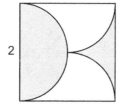

Problem 9.5 Find the areas of the shaded regions in each of the diagrams.

(a)

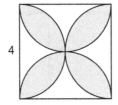

(b)

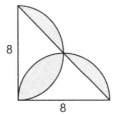

Problem 9.6 In the diagram below, the two quartercircles have radii 1 and 2 respectively. Find the difference between the areas of the two shaded regions.

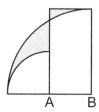

Problem 9.7 In the diagram, given that the side length of the square is 4. Find the total area of the whole shape.

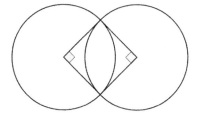

Problem 9.8 Suppose you have a circle of radius 1. A rectangle is inscribed in the circle, and a rhombus is inscribed in the rectangle, as shown in the diagram. What is the side length of the rhombus?

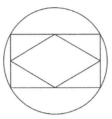

Problem 9.9 In the diagram, the largest circle has radius 5, and the two smaller circles have radii 3 and 4 respectively. The region A is the overlapped region of the two smaller circles. Find the difference between the area of the shaded region and the area of region A.

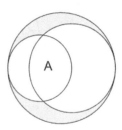

Problem 9.10 In the diagram, the area of region A equals the area of region B plus $50\pi - 100$. Find the height of the triangle.

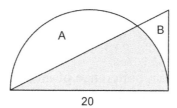

9.2 Quick Response Questions

Problem 9.11 What is the area of a circle with radius 5? Use $\pi = 3.14$. Round to the nearest tenth if necessary.

Problem 9.12 What is the circumference of a circle with radius 5? Use $\pi = 3.14$. Round to the nearest tenth if necessary.

Problem 9.13 In the following diagram, circle O has a radius length of 3 cm. and a ring with thickness measure 1 cm is placed around circle O 1 cm. apart. Is the area of circle O greater than the area of the ring?

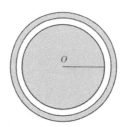

Problem 9.14 Find the circumference of a circle with area 16π. Use $\pi = 3.14$. Round to the nearest tenth if necessary.

Problem 9.15 Find the area of a circle with circumference 16π. Use $\pi = 3.14$. Round to the nearest tenth if necessary.

Problem 9.16 Tom and Jerry each eat some pizza. Tom's eats a quarter of his pizza, which has a radius of 10 inches. Jerry eats all of his pizza, which has a diameter of 10 inches. Is it true that Tom ate more crust than Jerry?

Problem 9.17 Tom and Jerry each eat some pizza. Tom's eats a quarter of his pizza, which has a radius of 10 inches. Jerry eats all of his pizza, which has a diameter of 10 inches. Is it true that Tom ate more pizza than Jerry?

Problem 9.18 What is the area of a sector of a 100° of a circle of radius 5? Use $\pi = 3.14$ and round your answer to the nearest tenth if necessary.

Problem 9.19 In the diagram, the square has side 8.

What is the area of the shaded region? Use $\pi = 3.14$ and round your answer to the nearest hundredth if necessary.

Problem 9.20 What is the length of an arc of 120° on a circle of radius 12? Use $\pi = 3.14$ and round your answer to the nearest tenth if necessary.

9.3 Practice Questions

Problem 9.21 Given a semicircle, let $\overline{AB}$ be its diameter, and O be the center. Let the radius be 4. Randomly select a point C on the arc. What is the maximum possible area of $\triangle ABC$?

Problem 9.22 Find the areas of the shaded regions in each of the diagrams. Note: for the most part things are drawn to scale, so you may assume angles that look right are in fact $90°$, etc.

(a)

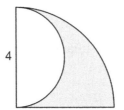

(b)

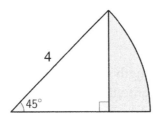

Problem 9.23 Find the areas of the shaded regions in each of the diagrams.

(a)

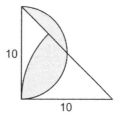

(b)

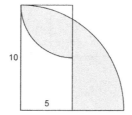

Problem 9.24 Find the area of the shaded region below.

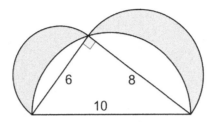

Problem 9.25 Find the area of the shaded region below.

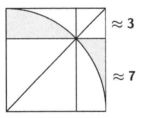

Problem 9.26 In the diagram below, the area of the shaded region is 4. What is the area of the semicircle with diameter $\overline{OA}$?

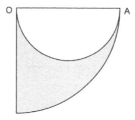

Problem 9.27 Suppose you have a triangle and a semicircle as in the diagram below. Find the difference between the area of region A and region B.

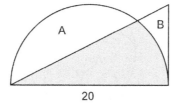

Problem 9.28 In the diagram, *ABCD* is a parallelogram, *O* is the center of the circle. Given that [*ABCD*] = 8, find the area of the shaded region △*BOC*.

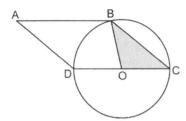

Problem 9.29 All the smaller circles in the diagram have radii 1. Find the perimeter of the shaded region.

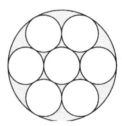

Problem 9.30 Find the perimeter of the shaded regions in each of the diagrams. Note: for the most part things are drawn to scale, so you may assume angles that look right are in fact 90°, etc.

(a)

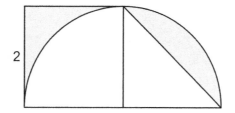

(b)

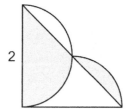

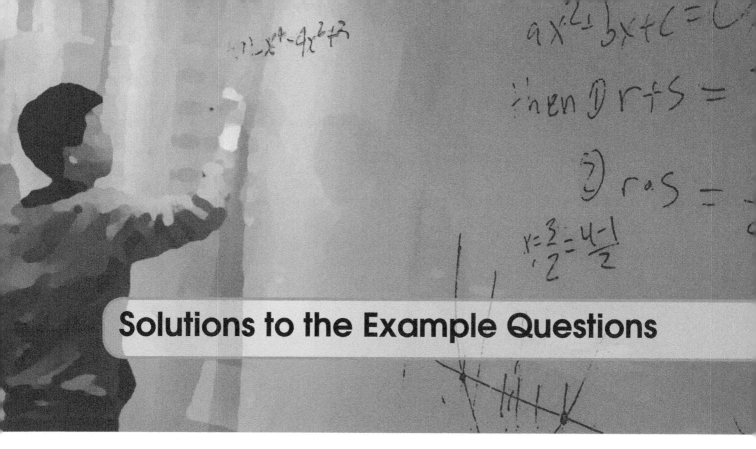

Solutions to the Example Questions

In the sections below you will find solutions to all of the Example Questions contained in this book.

Quick Response and Practice questions are meant to be used for homework, so their answers and solutions are not included. Teachers or math coaches may contact Areteem at info@areteem.org for answer keys and options for purchasing a Teachers' Edition of the course.

1   Solutions to Chapter 1 Examples

Problem 1.1 How many angles less than $180°$ are there in the following diagram?

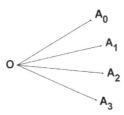

6.

Solution

There are 3 small angles ($\angle A_0OA_1$, $\angle A_1OA_2$, $\angle A_2OA_3$), 2 medium angles ($\angle A_0OA_2$, $\angle A_1OA_3$), and 1 large angle ($\angle A_0OA_3$). Therefore, the number of angles in the diagram is $3+2+1=6$.

Problem 1.2 In the following diagram, $\angle 1 = \angle 2 = \angle 3$. The sum of the measures of all possible angles in $\angle AOB$ is $180°$. What is the measure of $\angle AOB$?

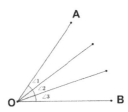

Answer

$54°$.

Solution

There are 6 angles in total. When adding up all the angles, $\angle 1$ and $\angle 3$ are counted 3 times each, and $\angle 2$ is counted 4 times. Therefore $3\angle 1 + 4\angle 2 + 3\angle 3 = 180°$. Since $\angle 1 = \angle 2 = \angle 3$, we get $10\angle 1 = 180°$, thus $\angle 1 = 18°$. So $\angle AOB = 3\angle 1 = 54°$.

Problem 1.3 Count the triangles in each of the following diagrams.

(a)

Answer

8

Solution

To count accurately, categorize the triangles. There are 4 one-piece triangles and 4 two-piece triangles, for a total of 8 triangles.

(b)

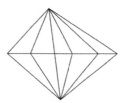

Answer

35

Solution

To count accurately, categorize the triangles. The top half of the diagram has $5 + 4 + 3 + 2 + 1 = 15$ triangles. Similarly the bottom has 15. Lastly there are 5 triangles which overlap the top and bottom for a total of 35 triangles.

Problem 1.4 In the following diagram, each small equilateral triangle has area 1. Find the total area of all the triangles in the figure below.

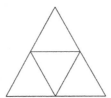

Answer

8.

Solution

In the diagram, there are 4 small triangles and 1 large triangle. Each small equilateral triangle has area of 1. Since the large triangle is made up of 4 small triangles, the area of the large triangle is 4. Therefore, the total area of all triangles is $4 \cdot 1 + 1 \cdot 4 = 8$.

Problem 1.5 How many squares are there in the following diagram?

Answer

5.

Solution

There are 4 small squares and 1 big square. The total number of squares is $4 + 1 = 5$.

Problem 1.6 In the following diagram, $ABCD$ is a parallelogram, and each of the segments in the diagram is parallel to one of $\overline{AB}$, $\overline{AD}$, or $\overline{BE}$. Count the number of parallelograms in the diagram that contain the shaded triangle.

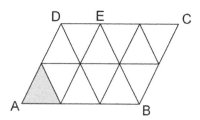

6.

Solution

If we ignore the top half of the diagram, we can create 3 parallelograms (made up of 2, 4, and 6 triangles). Including the top half creates an additional 3 parallelograms (made up of 4, 8, and 12 triangles). There are $3 + 3 = 6$ parallelograms containing the shaded region in the diagram.

Problem 1.7 Arrange several equilateral triangles, all of whose side lengths are 2cm, to form a long parallelogram, as shown in the diagram. Assume the perimeter of the long parallelogram is 144cm, how many triangles are there?

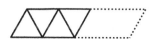

Answer

70.

Solution

The left and right sides each has length 2cm, so the remaining sides of the parallelogram have length 140cm. As each triangle has a side length of 2cm, there are $140/2 = 70$ triangles.

Problem 1.8 Given a circular disk, use 3 lines to divide the disk into small regions. At most how many regions can there be? What if there are 4 lines?

Answer

7 and 11.

Solution

One line makes 2 regions. Two lines make 4 regions. If a third line intersects the previous two lines, three new regions are created for a total of 7. Similarly, adding a fourth line that intersects the previous three will result in a total of 11 regions.

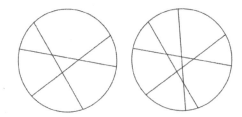

Problem 1.9 In the diagram, each side is perpendicular to its adjacent sides, and all small sides have equal length. Given that the perimeter of this diagram is 108cm, find the area of the shape.

Answer

405cm^2.

Solution

Note that the diagram has the same perimeter as a square with the same dimensions. The square thus has a side length of $108/4 = 27$cm. As a side of the square is made up of 3 small sides from the diagram, each small side has length $27/3 = 9$cm. As the diagram is made up of five squares with this side length, the total area of the shape is $5 \cdot 9^2 = 405$cm^2.

Problem 1.10 Use 4 congruent rectangles to form one big square, as shown. The big square has area 100 cm^2. Suppose the width of each rectangle is 1cm. What is the perimeter of each rectangle?

Answer

20cm.

Solution

The side length of the square is 10cm, so if the width of each rectangle is 1cm, the length must be 9cm. Hence the perimeter is $2 \cdot 1 + 2 \cdot 9 = 20$cm.

2 Solutions to Chapter 2 Examples

Problem 2.1 Two rectangles and two squares are assembled to form a big square as shown. The area of each rectangle is 28 and the area of the small square is 16. What is the area of the entire square?

Answer

121.

Solution

The little square's area is 16, so its side length is 4. Thus the small rectangle is 4×7, and therefore the dimensions of the entire square is 11×11, hence has area 121.

Problem 2.2 A rectangle is divided into 3 squares, as shown in the diagram. Given that the area of one bigger square is 12 in^2 more than that of one smaller square, find the area of the whole rectangle.

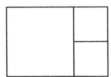

Answer

24 in^2.

Solution 1

Let a be the side length of the bigger square, b be the side length of the smaller square, then $a = 2b$, and $a^2 = b^2 + 12$. We have $a = 2b$, so $a^2 = 4b^2 = b^2 + 12$, thus $3b^2 = 12$, and $b^2 = 4$, so $a^2 = 16$. The area of the rectangle is $a^2 + 2b^2 = 24$ in^2.

Solution 2

Note the larger square must be made up of four of the small square. Hence the area of three small squares must be 12, and thus the entire square (equal to six small squares) must be 24.

Problem 2.3 A rectangle is divided into 3 squares, as shown in the diagram. Given that the area of the rectangle is 150 in^2. Find the length and width of the rectangle.

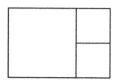

Answer

10 in and 15 in.

Solution 1

Let a be the side length of the bigger square, b be the side length of the smaller square, then $a = 2b$, and $a^2 + 2b^2 = 150$. We have $a = 2b$, so $a^2 = 4b^2$, so $a^2 + 2b^2 = 4b^2 + 2b^2 = 6b^2 = 150$. Hence, $b^2 = 25$ and $b = 5$, so $a = 10$. The width of the rectangle is 10in, and the length is $10 + 5 = 15$in.

Solution 2

Note the larger square must be made up of four of the small square, and thus the rectangle is made up of 6 small squares. Therefore, the small square has area $150/6 = 25$in^2 and thus the small square has side length 5in. Therefore the entire rectangle is $5 \cdot 3 = 15$ by $5 \cdot 2 = 10$.

Problem 2.4 A big rectangle is divided into 6 squares of different sizes, as shown. Given that the smallest square in the middle has area 4 cm^2 and the length of the big rectangle is 26, find the area of the big rectangle.

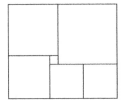

Answer

$572\,\mathrm{cm}^2$.

Solution

The smallest square has a side length of $2cm$. Let x be the side length of the square at the lower right corner. Then the side lengths of the squares (except for the smallest one), in clockwise order, are x, x, $x+2$, $x+4$, $x+6$. If the length of the rectangle is 26, we have $x+x+x+2 = 26$ (or also $x+4+x+6 = 26$), so $3x = 24$ and $x = 8$. We then see that the rectangle has width $8+14 = 22cm$. Hence the area is $22 \cdot 26 = 572cm^2$.

Problem 2.5 A big rectangle is divided into 7 smaller congruent rectangles, as shown. Given that the area of the big rectangle is $42\,\mathrm{cm}^2$, find the perimeter of the big rectangle.

Answer

26 cm.

Solution

Each smaller rectangle has area $6\,\mathrm{cm}^2$, and the ratio between the length and width of the smaller rectangle is $3 : 2$ (compare the top row of 2 horizontal rectangles and the middle row of 3 vertical ones). Therefore we can check that dimensions of 3 by 2 work for the small rectangle. The entire perimeter is thus $6 \cdot 3 + 4 \cdot 2 = 26cm$.

Problem 2.6 A rectangle is divided into 4 smaller rectangles by two lines, as shown. The perimeters of three of these rectangles are 12, 14, and 14. Find the perimeter of the remaining (shaded) rectangle.

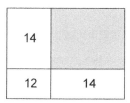

Answer

16.

Solution

Let x denote the length of the lower left rectangle. If its perimeter is 12, it must have a height of $6 - x$. Similarly, the upper left rectangle has dimensions x by $7 - x$. As the lower right rectangle has the same height as the lower left rectangle, its dimensions must be $x + 1$ by $6 - x$. Therefore, the dimensions of the shaded rectangle are $x + 1$ by $7 - x$ hence its perimeter is 16.

Problem 2.7 The perimeter of rectangle $ABCD$ is 20 cm. Construct a square on the top and right sides of $ABCD$ as shown below. Given that the sum of the areas of these squares is 52 cm^2, find the area of rectangle $ABCD$.

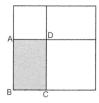

Answer

24 cm^2.

Solution

First note that the entire diagram is a square with dimensions 10×10, made up of the two squares and two congruent rectangles (one of which is $ABCD$). Hence the area of the square is $100 = 52 + 2[ABCD]$. Therefore, $[ABCD] = (100 - 52)/2 = 24$.

Problem 2.8 The shape in the diagram consists of 2 congruent squares and 3 congruent rectangles, and its perimeter is 14. Also given that $BC = \frac{1}{2}AB$. Find the area of rectangle $ABCD$.

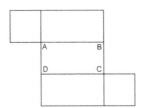

Answer

2.

Solution

Let x be the side length of the squares. Thus the rectangles are all $2x$ by $1x$. The perimeter is thus $14 = 10 \cdot 1x + 2 \cdot 2x = 14x$ and hence $x = 1$. Therefore the recangle is 2×1 with area 2.

Problem 2.9 Four congruent rectangles and one square are assembled into one big square. The areas of the two squares are 144 and 16 respectively. What are the length and width of the rectangles?

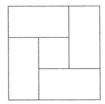

Answer

8 and 4.

Solution

The side lengths of the squares are 12 and 4 respectively. Therefore, the width of the rectangle is $(12-4)/2 = 4$, and the length of the rectangle is $(12+4)/2 = 8$.

Problem 2.10 Divide a big square into 6 congruent rectangles, as shown. Given that each of the rectangles has perimeter 140, find the area of the big square.

Answer

3600.

Solution

Let a be the length of each of the rectangles, and b be the width. Then $a + b = 70$, and $a = 6b$. Solve to get $a = 60, b = 10$. Therefore the side length of the square is $a = 60$, and finally the area of the big square is $60^2 = 3600$.

3   Solutions to Chapter 3 Examples

Problem 3.1 Consider the following diagram of two parallel lines and two transversals which meet at point C.

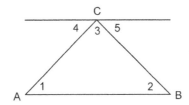

Which angles are equal in the above figure? Why?

Answer

$\angle 1 = \angle 4$ and $\angle 2 = \angle 5$

Solution

Since the two lines are parallel, by alternating interior angles, $\angle 1 = \angle 4$ and $\angle 2 = \angle 5$.

Problem 3.2 Consider the previous diagram of two parallel lines and two transversals which meet at point C. What is $\angle 3 + \angle 4 + \angle 5$?

Answer

$180°$

Solution

Since all of the angles are formed by a straight line, the sum of the angles is $180°$.

Problem 3.3 Consider the previous diagram of two parallel lines and two transversals which meet at point C. What is the sum of the angles in triangle ABC? Why?

Answer

$180°$

Solution

By the previous problems, since $\angle 3 + \angle 4 + \angle 5 = 180°$ and $\angle 1 = \angle 4$ and $\angle 2 = \angle 5$, $\angle 1 + \angle 2 + \angle 3 = 180°$.

Problem 3.4 Suppose that in the diagram below we have parallel lines and a transversal.

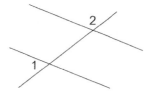

If the measure of $\angle 1$ is half the measure of $\angle 2$, find the measure of $\angle 1$.

Answer

$60°$.

Solution

Angles $\angle 1, \angle 2$ are same-side exterior angles, hence they are supplementary. As $\angle 2 = 2 \cdot \angle 1$, we have $\angle 1 + \angle 2 = 3 \cdot \angle 1 = 180°$ and hence $\angle 1 = 60°$.

Problem 3.5 Consider the diagram below:

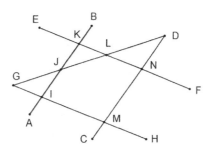

Suppose we know that $\overleftrightarrow{AB}$ and $\overleftrightarrow{CD}$ are parallel, $\angle DMH = 75°$, $\angle ELD = 140°$, and $\angle AJG = 35°$. What is $\angle LND$?

Answer

105°

Solution

We first have $\angle DLN = 180 - 140 = 40°$ (adjacent). Since $\overline{AB} \| \overline{CD}$, $\angle NDL = \angle AJG = 35°$ (corresponding). Therefore, as the angles in triangle $\triangle DNL$ add up to $180°$, $\angle LND = 180 - 40 - 35 = 105°$.

Problem 3.6 Consider the following "star" diagram, not drawn to scale.

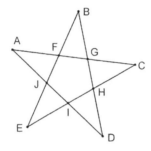

Suppose $\angle EFC = 120°$, $\angle AIC = 100°$, and $\angle BEC = 35°$. Calculate $\angle CAD$.

Answer

55°.

Solution

From adjacent angles we know $\angle AFJ = 60$ and $\angle AIE = 80$. Using $\triangle JEI$ we calculate $\angle EJI = 180 - 35 - 80 = 65°$. Hence $\angle AJF = \angle EJI = 65°$ (vertical angles). Hence $\angle CAD = \angle JAF = 180 - 65 - 60 = 55°$.

Problem 3.7 Given the figure below, if $\angle DFE = 60°$ and $\angle BCF = 90°$, what is the measure of $\angle CAF$?

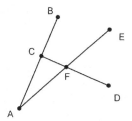

Answer

$30°$

Solution

Given $\angle DFE = 60°$, we observe that by vertical angles, $\angle AFC = 60°$. Given $\angle BCF = 90°$, we get $\angle FCA = 180° - 90° = 90°$. Therefore, $\angle AFC = 180° - 90° - 60° = 30°$.

Problem 3.8 Suppose $\angle AOF = 180°$ and is divided into five equal angles as shown below.

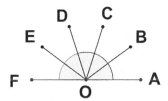

If $\angle AOB = \angle BOC = \cdots = \angle EOF$, find $\angle BOE$.

Answer

$108°$

Solution

All five angles are equal so each angle is $180/5 = 36°$. Since $\angle BOE$ contains three of the smaller angles, $\angle BOE = 108°$.

Problem 3.9 Consider the diagram below, where *l* and *m* are parallel but the drawing is not necessarily to scale.

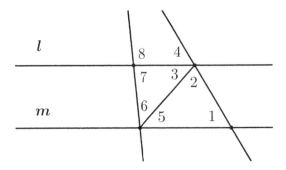

Suppose that $\angle 1 = 60°, \angle 5 = 50°, \angle 8 = 105°$. Find the measure of $\angle 4$.

Answer

$60°$

Solution

Since *l* and *m* are parallel, by corresponding angles, $60° = \angle 1 = \angle 4$. Therefore, $\angle 4 = 60°$.

Problem 3.10 Consider the diagram again.

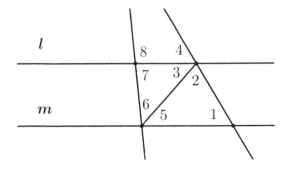

Suppose that $\angle 1 = 60°, \angle 5 = 50°, \angle 8 = 105°$. Find the measure of $\angle 2$.

Answer

$70°$

Solution

Since $\angle 1 = 60°$ and $\angle 5 = 50$, $\angle 2 = 180° - 60° - 50° = 70°$.

4   Solutions to Chapter 4 Examples

Problem 4.1 For each of the following "rules", state whether they work for proving congruence, similarity, both, or neither. If the rule does not work, give a counterexample.

(a) SAS (two sides and the angle between them)

Answer

Works for both congruence and similarity

(b) AAA (all three angles)

Answer

Works for similarity but not congruence

Solution

Note scaling a triangle (making it larger or smaller) does not change the angles.

(c) ASA (two angles and the side between them)

Answer

Works for both

Solution

Note that once we know two angles we know the third, so this could be referred to as AAA plus we know a side. We already know the triangles are similar (just from AAA), and knowing a side ensures congruence as well.

(d) AAS (two angles and a side not between them)

Answer

Works for both

Solution

Note that once we know two angles we know the third, so this could be referred to as AAA plus we know a side. We already know the triangles are similar (just from AAA), and knowing a side ensures congruence as well.

(e) SSA (two sides and an angle not between them)

Answer

Works for neither

Solution

Counterexamples may vary, for example cutting an isosceles triangle into two unequal pieces (through the third vertex).

Problem 4.2 Suppose $\triangle ABC$ is an isosceles triangle with $\angle A = \angle B$. Let D be a point on $\overline{AB}$.

(a) Prove that D is the midpoint of $\overline{AB}$ (we call $\overline{CD}$ the *median* from C) if and only if $\angle ACD = \angle BCD$ (we call $\overline{CD}$ the *angle bisector* of $\angle ACB$).

Solution

Remember we have two directions to prove!

First suppose D is the midpoint. We want to show $\angle ACD = \angle BCD$. Since the triangle is isosceles, $AC = BC$. By the definition of midpoint, $AD = BD$. As $CD = CD$, we can use SSS to show $\triangle ACD \cong \triangle BCD$. Hence $\angle ACD = \angle BCD$.

Now suppose $\angle ACD = \angle BCD$ and we want to show D is the midpoint of $\overline{AB}$. Identical to above we have $AC = BC$ and $CD = CD$. Therefore, we can use SAS to show $\triangle ACD \cong \triangle BCD$. Hence $AD = BD$ so D is the midpoint.

(b) Further prove that $\overline{CD}$ is perpendicular to $\overline{AB}$ (we call $\overline{CD}$ the *altitude* from C).

Solution

$\angle ACD$ and $\angle BCD$ are adjacent, so $\angle ACD + \angle BCD = 180°$. As they are equal, both must be $90°$.

Problem 4.3 Let $\overline{AD}$ and $\overline{BE}$ be medians in $\triangle ABC$. Prove that DE is half of AB.

First note that $\triangle DCE$ and $\triangle BCA$ share $\angle C$. Further, using the definition of midpoints we have $CB = 2 \cdot CD$ and $CA = 2 \cdot CE$. Hence, $\triangle DCE \sim \triangle BCA$ using SAS. As all the ratios in similar triangles are the same, $AB = 2 \cdot ED$ as needed.

Problem 4.4 Prove that the diagonals in a square are perpendicular.

Consider the labeling below:

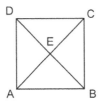

As a square is made up of four equal sides and four equal angles, using SAS we have $\triangle ABC \cong \triangle BCD \cong \triangle CDA \cong \triangle DAB$. Further, all of these triangles are isosceles, and hence must have angles 45-45-90. Hence, $\angle ABE = \angle BAE = 45°$, and so looking at triangle $\triangle ABE$ it is also a 45-45-90 triangle, so the diagonals are perpendicular as needed.

Problem 4.5 Prove that in a quadrilateral $ABCD$, if $\overline{AB}$ is parallel to $\overline{CD}$ and $\overline{BC}$ is parallel to $\overline{AD}$, then $AB = CD$ (and similarly $BC = AD$).

Draw diagonal BD. Using alternate interior angles, we have $\angle ABD = \angle BDC$, and $\angle ADB = \angle CBD$. Hence by ASA $\triangle BDA \cong \triangle DBC$ and $AB = CD$.

Problem 4.6 Let D, E be midpoints of $\overline{AB}$ and $\overline{BC}$ in $\triangle ABC$. Let F be the intersection of the perpendicular bisectors of $\overline{AB}$ and $\overline{BC}$. Show that AF, BF, CF all have the same length.

Solution

Note that using SAS we have $\triangle ADF \cong \triangle BDF$ (why?) and similarly $\triangle BEF \cong CEF$. Therefore we have $AF = BF$ and $BF = CF$ so all three have the same length.

Problem 4.7 Possible Triangles: Answer the following questions, with an explanation!

(a) Is it true that any three points are the vertices of some triangle?

Answer

No

Solution

If all three points are in a straight line, we do not get a triangle. Note this is sometimes called a *degenerate* triangle. However, if they are not all in a straight line, there is exactly one triangle with the three points as vertices.

(b) Given three numbers $p \leq q \leq r$ such that $p + q + r = 180$, is there a triangle $\triangle ABC$ such that $\angle A = p, \angle B = q, \angle C = r$? Hint: Suppose you make sure just $\angle A = p$ and $\angle B = q$, do you automatically know angle $\angle C = r$?

Answer

Yes, there are many

Solution

Start with a base $\overline{AB}$ and draw angles of $p°$ and $q°$ as in the diagram below:

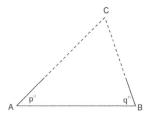

Let C be the intersection of the rays from the two angles. Note then in $\triangle ABC$, $\angle A =$

$p, \angle B = q, \angle C = 180 - p - q = r$ as needed. Further, if we start with a different size for the base, we get a different (non-congruent) triangle.

(c) Given three numbers $a \le b \le c$, is there a triangle $\triangle ABC$ with side lengths a, b, c? What can go wrong? Can you come up with a rule for when you can create a triangle with sides lengths a, b, c?

Answer

$c < a + b$.

Solution

We just need to ensure that the two shortest sides are long enough to reach the third. As long as they are long enough, we can put the sides together to make a triangle.

Problem 4.8 Suppose the two heights outside an obtuse triangle are the same length. Prove that the triangle is isosceles.

Solution

Let the triangle be $\triangle ABC$ with heights $BD = CE$ as shown.

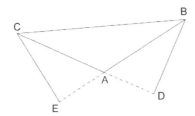

We have $BD = CE$ as a given. Both $\angle BDA, \angle CEA$ are right angles and hence equal. Lastly, $\angle CAE = \angle BAD$ (vertical angles). Therefore, by AAS, $\triangle CAE \cong \triangle BAD$ and hence $CA = BA$ as needed.

Problem 4.9 Suppose we have a "star" diagram as below (do not assume it is drawn to scale).

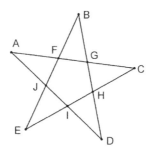

Now suppose that $\angle A = 30°$ and that $\triangle AFJ$ is isosceles. Calculate $\angle B + \angle D$.

Answer

$75°$.

Solution

We know $\triangle AFJ$ is isosceles, so $\angle AFJ = \angle AJF = (180 - 30)/2 = 75°$. Hence $\angle FJI = 180 - 75 = 105°$ (adjacent). Noting that the angles in $\triangle BJD$ must add up to $180°$, we have $\angle B + \angle D = 180 - 105 = 75°$.

Problem 4.10 Prove that if you connect the midpoints of the sides of an equilateral triangle it divides the triangle into four smaller congruent equilateral triangles.

Solution

Since all the side lengths of the triangle are equal, we can use SAS to show the three outer triangles are all congruent. Further, as they are isosceles with a $60°$ angle, they in fact are all equilateral triangles. Lastly, SSS shows that the inner triangle is congruent to the outer ones.

5   Solutions to Chapter 5 Examples

Problem 5.1 Prove the Pythagorean Theorem using the diagram below:

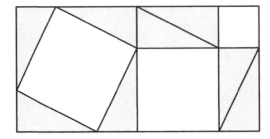

Solution

Let the triangles be denoted T, and the small, middle, and big squares denoted (respectively) S, M, B as labeled in the diagram below:

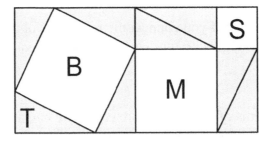

Note the entire diagram is made up of two congruent squares, so they have same area. The left square is $B + 4T$ and the right square is $S + M + 4T$. Hence, the area of square B is the sum of the squares M, S. If T has side lengths $a < b < c$, it is easy to see the areas of S, M, L are respectively a^2, b^2, c^2 and the result follows.

Problem 5.2 Prove the converse to the Pythagorean Theorem.

Solution

Suppose we have triangle $\triangle ABC$ with $AB = c, AC = b, BC = a$, if $c^2 = a^2 + b^2$ then $\triangle ABC$ is a right triangle. Now form a right triangle $\triangle A'B'C'$ with $A'C' = b$ and $B'C' = a$.

By the Pythagorean Theorem, $A'B' = c^2$. Hence, using SSS, $\triangle ABC \cong \triangle A'B'C'$ and therefore $\angle C'$ is right.

Problem 5.3 For each of the following, state whether it is possible to have a right triangle with the given side lengths. If it is possible, we call (a, b, c) a *Pythagorean Triple*.

(a) $3, 4, 5$.

Answer

Yes.

(b) $4, 5, 6$.

Answer

No.

(c) $5, 12, 13$.

Answer

Yes.

(d) $6, 8, 10$.

Answer

Yes.

(e) $5, 7, 8$.

Answer

No.

Problem 5.4 Suppose you have a right triangle $\triangle ABC$ with hypotenuse $AC = 13$. Attach right triangle $\triangle BCD$ with hypotenuse BC to the side of $\triangle ABC$. If $\triangle BCD$ has sides of length $3, 4$, find AB.

Answer

12.

Solution

Using the Pythagorean theorem $BC^2 = BD^2 + BC^2$, so $BC = 5$ (recall $(3,4,5)$ is a Pythagorean triple). Using the Pythagorean theorem again $AC^2 = AB^2 + BC^2$, so $AB = 12$ (as $(5,12,13)$ is another Pythagorean triple)

Problem 5.5 Let $ABCD$ be a rectangle with $AB = 6, BC = 3$. Let E be the point a third of the way from A to B on $\overline{AB}$. Is $\angle CED$ a right angle?

Answer

No.

Solution

Since E is a third of the way from A to B, we have $AE = 2$ and $BE = 4$. Therefore, we can use the Pythagorean Theorem on right triangles $\triangle BCE$ and $\triangle ADE$ to get that $CE^2 = 25$ and $DE^2 = 13$. We then see $36 = DC^2 \neq DE^2 + CE^2 = 25 + 13 = 38$, so $\triangle CDE$ is not a right triangle and hence $\angle CED$ is not a right angle.

Problem 5.6 Let's work with equilateral triangles.

(a) The sides of an equilateral triangle are 4 cm long. How long is an altitude of this triangle? What are the angles of a right triangle created by drawing an altitude? How does the short side of this right triangle compare with the other two sides?

Solution

$2\sqrt{3}$ cm; 30-60-90; The short side is half the hypotenuse, and $1/\sqrt{3}$ of the other leg.

(b) The altitudes of an equilateral triangle all have length 3 cm. How long are its sides?

Answer

$2\sqrt{3}$ cm

(c) If the area of an equilateral triangle is $4\sqrt{3}$ square inches, how long are its sides?

Solution

The area formula of an equilateral triangle: $\dfrac{\sqrt{3}}{4}a^2$ where a is the side length. So the side length is 4.

Problem 5.7 Suppose that $ABCD$ is a square. Let point E be *outside* the square and that $\triangle CDE$ is an equilateral triangle (see the diagram). What is the measure of $\angle EAD$?

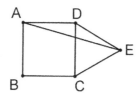

Answer

$15°$.

Solution

Since $DE = DC = DA$, $\triangle ADE$ is isosceles, therefore $\angle EAD = \angle DEA$. We know that $\angle ADE = 90° + 60° = 150°$, and $\angle ADE + \angle EAD + \angle DEA = 180°$, thus $\angle EAD = \dfrac{1}{2}(180° - 150°) = 15°$.

Problem 5.8 Given square $ABCD$, let P and Q be the points outside the square that make triangles CDP and BCQ equilateral. Segments $\overline{AQ}$ and $\overline{BP}$ intersect at G. Find angle AGP.

Answer

$90°$.

Solution

Identical arguments to previous problems give $\angle CBP = \angle BAQ = 15°$. Hence $\angle ABE = 90 - 15 = 75°$, and then $\angle BEA = 180 - 75 - 15 = 90°$ so $\angle AGP = 90°$ as well.

Problem 5.9 Given square $ABCD$, let P and Q be the points outside the square that make triangles CDP and BCQ equilateral. Prove that triangle APQ is also equilateral.

Solution

To prove $AP = PQ = QA$, we prove $\triangle ADP \cong \triangle QBA \cong \triangle QCP$ via SAS congruency. We have $\angle ADP = \angle QBA = \angle QCP = 150°$.

Problem 5.10 Draw the largest possible square inside an equilateral triangle, with one side of the square aligned with one side of the triangle. If the square has side length 6, find the side length of the equilateral triangle.

Answer

$6 + 4\sqrt{3}$.

Solution

Note that the square and equilateral triangle form two 30-60-90 triangles as in the diagram below

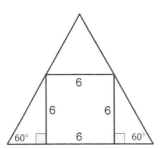

Note the bottom side of each 30-60-90 triangle is $6/\sqrt{3} = 2\sqrt{3}$, so the side length of the equilateral triangle is $6 + 2 \cdot 2\sqrt{3} = 6 + 4\sqrt{3}$.

6 Solutions to Chapter 6 Examples

Problem 6.1 Complete the following table about polygons: name, sum of interior angles, sum of exterior angles, and measure of each angle in case of regular polygon. All angles are in degrees. Justify your answers. Keep the chart for your own reference.

# of sides	Polygon name	Int. angle sum	Ext. angle sum	Each angle (if regular)
3	Triangle			
4				
5				
6				
7	Heptagon			
8				
10				
12	Dodecagon			

Solution

To justify the sum of interior angles of a triangle: In $\triangle ABC$, draw line through A that's parallel to $\overline{BC}$, then use the property of alternate interior angles. For polygons of more sides: cut them into triangles, and get the formula $(n-2)180°$.

For sum of exterior angles: Pretend the polygon's sides are streets, you are walking along the streets. Each time you turn a corner, you turn an exterior angle; when you return to the starting point, you turned a total of $360°$. You can also use the formula to calculate it. This gives the table below:

# of sides	Polygon name	Int. angle sum	Ext. angle sum	Each angle (if regular)
3	Triangle	180°	360°	60°
4	Quadrilateral	360°	360°	90°
5	Pentagon	540°	360°	108°
6	Hexagon	720°	360°	120°
7	Heptagon	900°	360°	$900/7°$
8	Octagon	1080°	360°	135°
10	Decagon	1440°	360°	144°
12	Dodecagon	1800°	360°	150°

Problem 6.2 Use 6 equilateral triangles to form a hexagon $ABCDEF$:

(a) Show hexagon $ABCDEF$ is regular. Why is this true?

Solution

It is clear that each side has the same length. Each angle in the hexagon is made up of two 60° angles (from the equilateral triangle), hence each angle is 120°.

(b) Calculate the angle AED.

Answer

90°.

Solution

Note that $\triangle AFE$ is isosceles with $\angle AFE = 120°$. Thus, $\angle FEA = (180 - 120)/2 = 30°$ and $\angle AED = \angle FED - \angle FEA = 120 - 30 = 90°$.

Problem 6.3 Three non-overlapping regular plane polygons all have sides of length 1.

The polygons meet at a point A in such a way that the sum of the three interior angles at A is $360°$. Among the three polygons, one is a triangle and one is a dodecagon. Find the remaining polygon.

Answer

Dodecagon.

Solution

The remaining angle is $360 - 60 - 150 = 150°$, which belongs to an another dodecagon (12 sides).

Problem 6.4 Let $ABCDEF$ be a regular hexagon, and let $EFGHI$ be a regular pentagon. Find all possible values of measure of $\angle GAF$.

Answer

$84°, 24°$

Solution

If the pentagon is inside the hexagon, then $\angle GAF = 84°$. If the pentagon is outside the hexagon, then $\angle GAF = 24°$.

Problem 6.5 Suppose that $DRONE$ is a regular pentagon, and that $DRUM$, $ROCK$, $ONLY$, $NEAP$, and $EDIT$ are squares attached to the outside of the pentagon. Is the decagon $ITAPLYCKUM$ equiangular? Is it equilateral?

Solution

It is equiangular, but not equilateral.

Problem 6.6 Explain where and how to cut off the sides of an equilateral triangle to form a regular hexagon.

Solution

If we place points dividing each side into thirds and draw parallel line segments we get the diagram below.

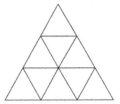

All of these triangles are congruent equilateral triangles (why?) and it is easy to see that removing the outer three results in a regular hexagon.

Problem 6.7 Practice with Hexagons

(a) Find the area of a regular hexagon with side length 12.

Answer

$216\sqrt{3}$.

Solution

The hexagon is made up of 6 equilateral triangles, each with an area of $\frac{1}{2} \cdot 12 \cdot 6\sqrt{3} = 36\sqrt{3}$. Hence the area is $6 \cdot 36\sqrt{3} = 216\sqrt{3}$.

(b) Find the side length of a regular hexagon with area $150\sqrt{3}$.

Answer

10.

Solution

The hexagon is made up of 6 equilateral triangles, each with an area of $25\sqrt{3}$. Therefore, the side length must be 10.

Problem 6.8 Mark Y inside regular pentagon $PQRST$, so that PQY is equilateral. Is RYT straight? Explain.

Answer

No.

Solution

We have $\angle PYT = \angle RYQ = 66°$, so $\angle RYT = 168°$.

Problem 6.9 Attach a regular pentagon $ABDEF$ to the side of an equilateral triangle ABC. Calculate $\angle CDE$.

Answer

$102°$.

Solution

Note $\triangle BCD$ is an isosceles triangle, and $\angle CBD = 60 + 108 = 168°$. Hence $\angle BDC = (180 - 168)/2 = 6°$, so $\angle CDE = 108 - 6 = 102°$.

Problem 6.10 A stop sign — a regular octagon — can be formed from a square sheet of metal by making four straight cuts that snip off the corners. If we want an octagon with sides of length $\sqrt{2}$, how large does the side of the original square need to be?

Answer

$2 + \sqrt{2}$.

Solution

Note the four corners cut off are each 45-45-90 triangles with hypotenuse $\sqrt{2}$. Hence their other sides have length 1. Hence, the side of the square is $\sqrt{2} + 1 + 1 = 2 + \sqrt{2}$.

7 Solutions to Chapter 7 Examples

Problem 7.1 Using only the basics about parallel lines and congruent/similar triangles and the fact that the area of the rectangle is bh, explain the following: (Note: once a fact is proven below, you can use it in later parts.

(a) The area of a parallelogram is bh.

Solution

As in the diagram below, cutting a triangle at one end of the parallelogram and moving it to the other results in a $b \times h$ rectangle.

(b) The area of a triangle is $\frac{1}{2}bh$ (prove this two ways!).

Solution

Method 1: A right triangle is half a rectangle, so has area $\frac{1}{2}bh$. Then any triangle can be split into two right triangles (by dropping an altitude to the longest side).

Method 2: Two copies of any triangle can be combined to form a parallelogram with base b and height h.

(c) The area of a trapezoid is $\frac{b_1+b_2}{2}h$.

Solution

Draw a diagonal of the trapezoid. This breaks the trapezoid into two triangles, each with height h and having bases b_1 and b_2.

Problem 7.2 Explain the following:

(a) Triangles (or parallelograms) with equal bases and equal heights have the same area.

Clear from the formulas.

(b) In $\triangle ABC$, let D be the midpoint of side $\overline{BC}$ and connect $\overline{AD}$, as in the diagram below.

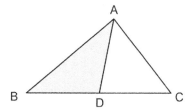

Prove that $[ABD] = [ACD]$ (that is, $\triangle ABD$ has the same area as $\triangle ACD$).

Solution

Note the triangles have the same height and the bases are equal as D is the midpoint.

(c) Let $ABCD$ be a parallelogram and E be any point on side $\overline{CD}$ as shown.

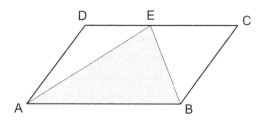

Prove that $[ABE] = \frac{1}{2}[ABCD]$.

Solution

Regardless of where E is on $\overline{CD}$, $\triangle ABE$ has the same height and base as the parallelogram.

(d) In $\triangle ABC$, let D be any point on side $\overline{BC}$ as shown below.

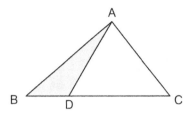

Prove that the ratios $\dfrac{[ABD]}{[ACD]} = \dfrac{BD}{CD}$.

Solution

Note both triangles share the same height, say h. Then $[ABD] = \frac{1}{2}h \cdot AB$, $[ACD] = \frac{1}{2}h \cdot CD$ and the result follows.

Problem 7.3 Given three segments of lengths 4cm, 6cm, and 8cm. Use these lengths as the bases and altitude (not necessarily in the given order) to make a trapezoid. Trapezoids of three different possible areas can be made. Which one is the largest, and what is its area?

Answer

40

Solution

Calculate the areas of trapezoids using each of the 3 lengths as altitude and the remaining two as bases. The areas are: If 5cm is the height: $(6+8) \times 4/2 = 28$; If 8cm is the height: $(4+8) \times 6/2 = 36$; If 10cm is the height: $(4+6) \times 8/2 = 40$. Thus the largest area is obtained by using the longest segment has the height.

Problem 7.4 The parallelogram $ABCD$ has area 48cm^2, $AE = 8$cm, $CE = 4$cm. Find the area of the shaded region.

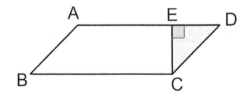

Answer

8cm^2

Solution

$AD = 48/4 = 12$, and $ED = 12 - 8 = 4$, so the area $[DEC] = \dfrac{1}{2}4 \cdot 4 = 8 \text{ cm}^2$.

Problem 7.5 A garden of rectangular shape is shown in the diagram. The shaded regions are grass, and the unshaded regions are empty spaces in the shape of four congruent rhombi. Find the ratio between the areas of the grass and empty regions.

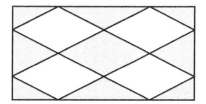

Answer

$1 : 1$.

Solution

Note that the rectangle consists of 4 unshaded rhombi, 1 shaded rhombi, 4 shaded half-rhombi, and 4 shaded quarter-rhombi. Further, each of these rhombi is congruent (why?). The ratio is thus $4 : (1 + 4 \cdot .5 + 4 \cdot .25) = 4 : 4$.

Problem 7.6 In the diagram, points A and B are the midpoints of their respective sides. Compute the ratio of the area of shaded region and the whole rectangle.

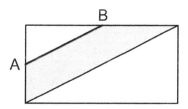

Answer

$3:8$.

Solution

Connect A and B and the midpoint of the diagonal, and it is clear that the shaded area is $3/4$ of the upper half of the rectangle. So the ratio to the whole rectangle is $3:8$.

Problem 7.7 In the given trapezoid $ABCD$, there are 8 triangles. Among them, the pair $\triangle ABC$ and $\triangle DBC$ have the same area. How many other pairs have the same areas? List the pairs with the same areas.

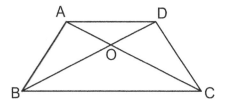

Answer

3 total pairs.

Solution

There are 2 more. $\triangle ABD, \triangle ACD$, and $\triangle ABO, \triangle DCO$.

Problem 7.8 Let $ABCD$ be a parallelogram with $[ABCD]$. Let P be a point in the interior of $ABCD$. Show that $[ABP] + [CDP] = [ABCD]/2$.

Answer

6.

Solution 1

Construct line segment $\overline{EF}$ with E, F on $\overline{AD}, \overline{BC}$ such that $\overline{EF}$ goes through P and $\overline{EF} \parallel \overline{AB}, \overline{CD}$. We then have: $[CDP] = [CDF]$ and $[ABP] = [ABF] = [DBF]$ (draw these out!). Hence, $[ABP] + [CDP] = [DBF] + [CDF] = [BCD] = [ABCD]/2$.

Solution 2

Note the height from $\overline{AB}$ to $\overline{CD}$ through P splits into altitudes for the two given triangles. As these altitudes sum to the total height of the parallelogram, $[ABP] + [CDP] = [ABCD]/2$.

Problem 7.9 Let $ABCD$ be a parallelogram, as in the diagram. Compare the shaded regions $\triangle ABF$ and $\triangle DEF$, which one has the larger area?

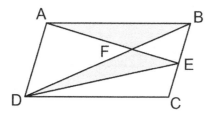

Answer

The have the same area.

Solution

Note $[ABD] = [DEF]$ as they have the same height and base. Hence, after removing the shared $\triangle AFD$ we see the two shaded triangles have the same area.

Problem 7.10 If regular hexagon $ABCDEF$ with side length 2 can be reinterpreted as 6 equilateral triangles, what is the area of $ABCDEF$?

Answer

$6\sqrt{3}$.

Solution

Note that $ABCDEF$ can be decomposed into 6 equilateral triangles each with side length 2. We can apply the rules of special right triangles to deduce that the height of one equilateral triangle is $\sqrt{3}$. The area of one equilateral triangle is $\frac{1}{2}(2)(\sqrt{3}) = \sqrt{3}$. Since there are six equilateral triangles in $ABCDEF$, $[ABCDEF] = 6\sqrt{3}$.

8   Solutions to Chapter 8 Examples

Problem 8.1 All the rectangles in the following diagrams are squares. The lengths of the segments are marked. Find the area of the shaded regions in each diagram.

(a)

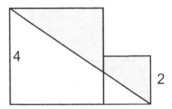

Answer

8

Solution

The shaded region is the two squares minus the unshaded triangle with base 6 and height 4, so the area is $4^2 + 2^2 - 4 \cdot 6/2 = 8$.

(b)

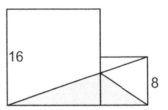

Answer

64

Solution

The shaded region is a triangle with base and height $24, 8$ minus a triangle with base and height $8, 8$, so the area is $24 \cdot 8/2 - 8 \cdot 8/2 = 64$.

(c)

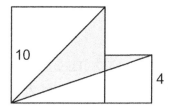

38

The shaded region is the two squares minus triangles with base and height $10, 10$ and $14, 4$, so the area is $10^2 + 4^2 - 10 \cdot 10/2 - 14 \cdot 4/2 = 38$.

(d)

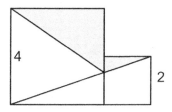

6

The shaded region is the two squares minus triangles with base and height $4, 4$ and $6, 2$, so the area is $4^2 + 2^2 - 4 \cdot 4/2 - 6 \cdot 2/2 = 6$.

Problem 8.2 Suppose $ABCD$ is a parallelogram, and the altitude $\overline{AE}$ on side $\overline{BC}$ is 5. Suppose the difference between the areas $[ADCE]$ and $[ABE]$ is 15. Find the length of $\overline{EC}$.

Answer

3

Solution

Draw $\overline{EF}$ parallel to $\overline{CD}$ with F on $\overline{AD}$. Then $\triangle ABE$ and $\triangle EFA$ are congruent, and they have the same area, therefore $[ECDF] = 15$. Thus $EC = 15/5 = 3$.

Problem 8.3 In the diagram below, there are 36 rectangular grid points, evenly spaced, and the distance between each pair of adjacent points is 1. Find the area of $\triangle ABC$.

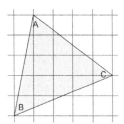

Answer

23/2

Solution

The square surrounding $\triangle ABC$ is 5×5 so has area 25. We subtract of the areas of the three triangles outside ABC: $25 - \frac{1}{2} \cdot 1 \cdot 5 - \frac{1}{2} \cdot 5 \cdot 2 - \frac{1}{2} \cdot 3 \cdot 4 = 23/2$.

Problem 8.4 In the diagram below, there are 21 grid points arranged in equilateral triangles, equally spaced. The *area* of each small equilateral triangle formed by 3 adjacent grid points is 1. Find the area of $\triangle ABC$.

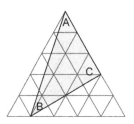

Answer

12

Solution

Call the entire triangle ADE (with C on $\overline{AE}$). $\triangle ABC, \triangle ABE$ have the same height, so $[ABC] : [ABE] = 3 : 5 = 12 : 20$. Similarly, $[ABE] : [ADE] = 4 : 5 = 20 : 25$. Combining these two we have $[ABC] : [ADE] = 12 : 25$. As $[ADE] = 25$, we have $[ABC] = 12$.

Problem 8.5 In the diagram, $\triangle ABC, \triangle DEF$ are two congruent isosceles right triangles. Given that $AB = 6, EC = 2$, find the area of the shaded region.

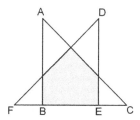

Answer

12

Solution

Label the resulting pentagon $BEGHI$. We have $AB = BC = 6$, so $BE = 6 - 2 = 4$. Therefore, $FB = 2$ so $CF = 8$. Now note $\triangle CFH$ is a 45-45-90 triangle with hypotenuse 8. Hence its sides are length $8/\sqrt{2}$, so it has area $[CFH] = \dfrac{1}{2} \cdot \dfrac{8}{\sqrt{2}} \cdot \dfrac{8}{\sqrt{2}} = 16$. Similarly, $[CEG] = [FBI] = \dfrac{1}{2} \cdot 2 \cdot 2 = 2$. Hence, the shaded pentagon has area $16 - 2 \cdot 2 = 12$.

Problem 8.6 Suppose $\triangle ABC$ with E on $\overline{AB}$ and D on $\overline{AC}$ such that $AE = AB/3, AD = AC/2$. If $[AED] = 2$, find the area of $[ABC]$.

Answer

12

Solution

We have $[ABD] = 3[AED]$ (same height) and similarly $[ABC] = 2[ABD]$, so $[ABC] = 6[AED] = 2 \cdot 6 = 12$.

Problem 8.7 Parallelogram $ABCD$ is shown below, where triangle BCE is a right isosceles triangle and A is the midpoint of $\overline{BE}$. Given that $[ABCF] - [DFC] = 4$, find the area of $ABCD$.

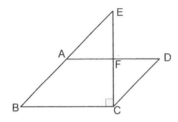

Answer

8

Solution

Since $\triangle BCE, \triangle AFE$ are 45-45-90 triangles and A is a midpoint, we get that F is the midpoint of $\overline{AD}$ (why?). From here we get that $[ABCF] = \frac{3}{4} \cdot [ABCD], [DFC] = \frac{1}{4} \cdot [ABCD]$, so $[ABCF] - [DFC] = \frac{1}{2} \cdot [ABCD]$. Hence $[ABCD] = 8$.

Problem 8.8 As shown in the diagram, square $ABCD$ has side length 5. Let E and F be the midpoints of $\overline{AB}$ and $\overline{BC}$ respectively. Find the area of quadrilateral $BFGE$.

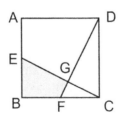

Answer

5

Solution

Connect the other two midpoints to vertices as shown:

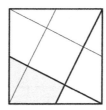

Note the piece can be rearranged to form 5 congruent squares. The area of the big square is the sum of the areas of the 5 small squares, so each of the 5 small squares has area $5^2/5 = 5$. The area of the shaded region is the same as one small square, so the answer is 5.

Problem 8.9 In the diagram, $\triangle ABC, \triangle DEF$ are two congruent isosceles right triangles. Given that $AB = 9, EC = 3$, find the area of the shaded region.

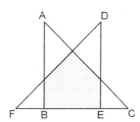

Answer

27

Solution

Label the resulting pentagon $BEGHI$. We have $AB = BC = 9$, so $BE = 9 - 3 = 6$. Therefore, $FB = 3$ so $CF = 12$. Now note $\triangle CFH$ is a 45-45-90 triangle with hypotenuse 12 (why?). Hence its sides are length $12/\sqrt{2}$, so it has area $[CFH] = \dfrac{1}{2} \cdot \dfrac{12}{\sqrt{2}} \cdot \dfrac{12}{\sqrt{2}} = 36$. Similarly, $[CEG] = [FBI] = \dfrac{1}{2} \cdot 3 \cdot 3 = \dfrac{9}{2}$. Hence, the shaded pentagon has area $36 - 9 = 27$.

Problem 8.10 In the figure containing right triangle BCE and parallelogram $ABCD$ shown below, $EF = 3$, $EG = 5$ and $BG = 20$. Find the area of the shaded region.

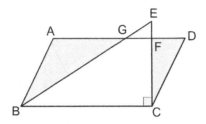

96

Given right triangle BCE, we have that EFG is also a right triangle. This implies that $FG = 4$ and $BC = 20$ by similar triangles. Therefore, $CF = 12$ by Pythagorean Theorem. The area of the shaded region is $(20)(12) - \dfrac{1}{2}(12)(20+4) = 96$.

9   Solutions to Chapter 9 Examples

Problem 9.1 A circular dining table has diameter 2 meters and height 1 meter. A square tablecloth is placed on the table, and the four corners of the tablecloth just touch the floor. Find the area of the tablecloth in square meters.

Answer

$8m^2$.

Solution

As shown in the diagram, the tablecloth's diagonal is 4 meters, thus the area is $4^2/2 = 8$ m^2.

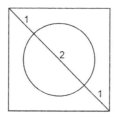

Problem 9.2 The shape in the diagram consists of three semicircles with radii 5, 3, and 2, arranged as shown. Calculate the perimeter and area of the shaded region.

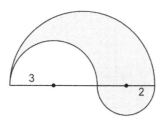

Answer

$10\pi, 10\pi$.

Solution

Perimeter $2\pi \cdot 3/2 + 2\pi \cdot 2/2 + 2\pi \cdot 5/2 = 10\pi$, and area $\dfrac{\pi 5^2}{2} - \dfrac{\pi 3^2}{2} + \dfrac{\pi 2^2}{2} = 10\pi$.

Problem 9.3 Find the areas of the shaded regions in each of the diagrams. Note: for the most part things are drawn to scale, so you may assume angles that look right are in fact 90°, etc.

(a)

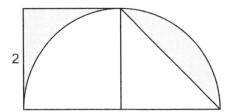

Answer

2

Solution

The two shaded regions combine to form half a 2 by 2 square. Thus the area is 2.

(b)

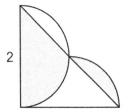

Answer

$\pi/2$

Solution

The two shaded regions combine to form a semicircle with radius 1. Thus the area is $\pi/2$.

Problem 9.4 Find the areas of the shaded regions in each of the diagrams. Note: for the most part things are drawn to scale, so you may assume angles that look right are in fact 90°, etc.

(a)

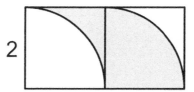

Answer

4

Solution

The two shaded regions combine to form a 2 by 2 square. Thus the area is 4.

(b)

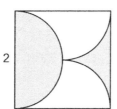

Answer

2

Solution

The two shaded regions combine to form a 2 by 1 rectangle. Thus the area is 2.

Problem 9.5 Find the areas of the shaded regions in each of the diagrams.

(a)

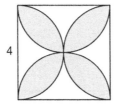

Answer

$8\pi - 16$

Solution

Start by drawing the diagonals. Note the 8 regions can be rearranged to form 2 circles (each with radius 2) so that a square (with diagonal 4) is left unshaded in the middle of each square. Hence, the shaded area is $2 \cdot \pi 2^2 - 2 \cdot 4^2 / 2 = 8\pi - 16$.

(b)

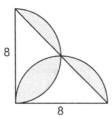

Answer

$16\pi - 32$

Solution

The shaded region can be broken down into 4 identical slices. These slices can be rearranged to get a circle of radius 4 with missing square of diagonal 8. Hence, the shaded area is $\pi 4^2 - 8^2 / 2 = 16\pi - 32$.

Problem 9.6 In the diagram below, the two quartercircles have radii 1 and 2 respectively. Find the difference between the areas of the two shaded regions.

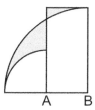

$3\pi/4 - 2$.

Solution

A quartercircle with radius 2, minus a quartercircle with radius 1, minus a rectangle with length 2 and width 1: $\dfrac{2^2\pi}{4} - \dfrac{1^2\pi}{4} - 2 \times 1 = \dfrac{3\pi}{4} - 2$.

Problem 9.7 In the diagram, given that the side length of the square is 4. Find the total area of the whole shape.

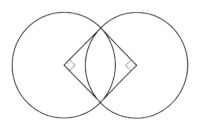

Answer

$24\pi + 16$.

Solution

Both circles have radii 4, and the desired area is the sum of the two circles minus their overlap region. The overlap region can be calculated by adding two quartercircles and then subtracting the square. Thus the overlap region of the two circles: $2 \cdot \dfrac{4^2\pi}{4} - 4^2 = 8\pi - 16$. Then the final answer is $2 \cdot 4^2\pi - (8\pi - 16) = 24\pi + 16$.

Problem 9.8 Suppose you have a circle of radius 1. A rectangle is inscribed in the circle, and a rhombus is inscribed in the rectangle, as shown in the diagram. What is the side length of the rhombus?

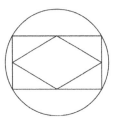

Answer

1.

Solution

The side length of the rhombus is the same as the radius of the circle. So the answer is 1.

Problem 9.9 In the diagram, the largest circle has radius 5, and the two smaller circles have radii 3 and 4 respectively. The region A is the overlapped region of the two smaller circles. Find the difference between the area of the shaded region and the area of region A.

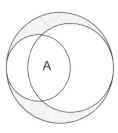

Answer

0.

Solution

Since $3^2 + 4^2 = 5^2$, the area of the largest circle equals the sum of the two smaller circles. Thus the region A and the shaded region have the same area. Therefore the answer is 0.

Problem 9.10 In the diagram, the area of region A equals the area of region B plus $50\pi - 100$. Find the height of the triangle.

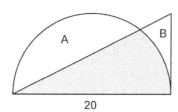

Answer

10.

Solution

Let the height be x. The area A minus the area B is equal to the semicircle minus the triangle: $\dfrac{10^2\pi}{2} - \dfrac{20x}{2} = 50\pi - 100$, and solve for x.